RF $3⁴⁰ U

MAN-APES
OR APE-MEN?

The Story of Discoveries in Africa

SIR WILFRID E. LE GROS CLARK
Emeritus, Oxford University

MAN-APES
OR APE-MEN?

The Story of Discoveries in Africa

HOLT, RINEHART AND WINSTON, INC.
NEW YORK · CHICAGO · SAN FRANCISCO · TORONTO · LONDON

PREFACE

In 1925 the announcement was made of the discovery in South Africa of the fossil skull of an apelike creature, a juvenile individual with a small brain of simian dimensions but with cranial and dental characters very different from those of modern apes and approximating to those distinctive of the human family (the Hominidae as this family is termed in zoological nomenclature). To this specimen the name *Australopithecus* or "southern ape" was given, and the group as a whole to which it belongs has come to be known as the australopithecines. In succeeding years very numerous remains of these creatures were discovered by Professor Raymond Dart, the late Dr. Robert Broom, and Professor J. T. Robinson at different sites in the Transvaal, and these discoveries have led to incessant controversies on the significance of the fossils. The main focus of argumentation was at first concentrated on the question whether the australopithecines are to be regarded as exceedingly primitive hominids with brains hardly larger than those of the modern big apes, or perhaps no more than apes that had, by a process of parallel evolution, developed certain features merely simulating those characteristic of the Hominidae. This preliminary controversy has now been finally settled, for there is general agreement that they were certainly primitive representatives of the hominid family and quite distinct from the ape family. But controversies have continued over the precise relationship of different australopithecine types whose remains have been found at various sites in South Africa and, more recently, in East Africa. Some of these controversies have, I think, been based on false premises and they have led to a good deal of confusion, particularly for those who have no intimate and firsthand acquaintance with the fossil material. It has seemed to me to be appropriate now to clarify the issues by emphasizing the need to recognize that the australopithecines comprised a widely ranging group consisting of local populations inhabiting Africa south of the Sahara in early times, from about two million to half a million years ago. It should be accepted, therefore, that the different local populations, separated in space as well as in time, are likely to have shown a considerable degree of variability in the minor details of their anatomy. But such a degree of variation does not appear to me to justify claims that some of these local populations are not assignable to the group of the australopithecines as a whole. If due account is taken of the variability to be expected in different local populations, a variability no greater than that found among the modern large apes or in the various types of *Homo* (modern and extinct), the whole problem of the evolutionary significance of the australopithecines becomes greatly simplified. In this book I have followed the sequence of controversies from 1925 to the present time, with the endeavor to resolve conflicting opinions by a review of the nature of the evidence on which they have been based.

W. E. LE GROS CLARK

Oxford, England
November 1966

CONTENTS

MAN-APES
OR APE-MEN?

The Story of Discoveries in Africa

1

MAN'S NEAREST
RELATIONS

Let me first explain the meaning of the title of this book, *Man-apes or Ape-men?* The term "man-ape" or "anthropoid ape" has long been applied to the tailless apes that are much more closely similar in their anatomical structure to man himself than to the monkeys. Those apes that still exist today include the gibbon, orangutan, gorilla, and chimpanzee. At one time they were grouped together in a single zoological category technically called the Anthropomorpha or Simiidae. Most authorities still agree that they constitute a single zoological family, but in accordance with an awkward nomenclatural priority this family is now, less aptly I think, termed the Pongidae. The reason for the change is that the first anthropoid ape to be described and given a scientific name of its own was an orangutan and it was given the generic name *Pongo*. To complete these technical terms for the moment, we may note that the Pongidae are distinguished on the one side from the Hominidae (that is, the human family), and on the other from all the types of monkey inhabiting the Old World, the Cercopithecidae. Over a hundred years ago, in 1863, the famous anatomist Thomas Henry Huxley made the statement: "Whatever system of organs be studied, the comparison of their modifications in the

1

ape series leads to one and the same result—that the structural differences which separate man from the gorilla and chimpanzee are not so great as those which separate the gorilla from the lower apes." By "lower apes" he was referring to what we now term monkeys.

All subsequent studies have served to amplify this declaration of Huxley's and, indeed, have reinforced it more and more emphatically; so much so, that in the zoological classification of the order of mammals called the primates the human and anthropoid ape families are now usually included in a common group, a superfamily called the Hominoidea (or, colloquially, hominoids). The evolutionary implication of this classification is simply this, that the two families have in the distant geological past been derived from a common ancestral stock which, in the course of time, became diversified in two directions, one leading to the modern and extinct anthropoid apes, and the other to the modern and extinct types of man. But this does not mean that the common ancestral stock would have shown anatomical characters precisely intermediate between those of the Pongidae and the Hominidae, for each of these families has developed adaptive specializations peculiar to itself. For example, the anthropoid apes, or pongids, show specializations in their limbs related to their habit of swinging among the branches of trees with their long arms, a mode of progression which has been called brachiation. On the other hand, the hominids have become specialized quite otherwise in developing their lower limbs and other parts of the body for bipedal locomotion in an erect posture. However, in so far as the pongids have preserved more primitive characteristics, such as the small size of their brain and their large jaws, it is reasonable to hypothesize that in the early stages of their progressive diversification from a common ancestral stock the forerunners of the modern hominids obviously would have been more apelike than the latter. In fact, fossil remains of such early hominids are now known, and, because they were representatives of the hominid line of evolution and yet were more apelike than *Homo sapiens* in some of the primitive features that they still at that time retained, they have sometimes been called colloquially "ape-men" in contradistinction to the "man-apes."

Now let us consider some of the indirect evidence that suggests a consanguinity and therefore, it is presumed, a close evolutionary relationship between man and the large anthropoid apes. In Darwin's time, and for many years afterward, this indirect evidence was based on the study of the comparative anatomy of the two groups; identity, or near identity, of structure was assumed, as in the case of other animals, to betoken a relationship in the genetic sense, and in the case of some mammals this assumption has received corroboration from fossil remains which actually demonstrated connecting links between different groups that show structural similarities. The large apes are certainly very "manlike" in many ways. This is apparent enough even to the unsophisticated eye and it is for that reason, of course, that they were called anthropoid or anthropomorphous apes. For one thing, on occasion they may be able to assume momentarily an upright, or orthograde, posture in standing and walking, unlike the habitual pronograde posture of monkeys. It is true that gorillas and chimpanzees do, in a fashion, commonly move about on all fours except when they are climbing, while gibbons and orangutans spend most of their active time swinging among the treetops. But all of them, like man, have a broad, flat chest, and a characteristic disposition of the internal visceral organs that are functionally related to the tendency to hold the trunk in a vertical rather than a horizontal position when sitting or climbing, and this is in strong contrast with monkeys in which the chest is narrowed from side to side and the viscera slung in position as in lower quadrupedal mammals. It was no doubt such a change of trunk posture which, so to speak, preconditioned the development of a completely upright stance in man. There are significant features in the pelvis also for in apes and man the tail has disappeared, leaving behind a few nodules of vertebrae that together form the coccyx.

Again, the brain of the large apes is astonishingly like the human brain—smaller, of course, but constructed on the same basic pattern with a similar, if simpler, pattern of convolutions of the gray matter of the cerebral cortex. The skull approximates more closely to the human skull than does that of the monkeys in the shortness of the face (relative to the size of the skull as a whole) and in the volume of the braincase. Some of the limb bones of a chimpanzee may be quite difficult to distinguish from human limb bones, for example the upper arm bone or humerus. The foot skeleton and the muscles associated with it, in spite of the divergent big toe, show many striking similarities to the hominid foot skeleton. In fact, there are very good reasons for supposing that the characteristic enlargement of the big toe in man could hardly have arisen as a product of evolution unless it had been derived from a large and powerful grasping big toe very similar to that of anthropoid apes [49]. The teeth of the large apes resemble in many respects those of man, and it is sometimes difficult to tell from an isolated molar tooth whether it is that of a chimpanzee or

Schuman & Bruce
'54 H.B. "Metric
morphologic var-
iations in the dentition
of the Siberian chimp"

of some type of man [75].[1] And although the long, pointed and overlap-
ping canine teeth of the apes differ markedly from the small canine of
modern man, we now know that in some of the extinct types of man that
lived many thousands of years ago the canine was much more strongly developed. We could also mention many details of their muscular anatomy
and the form and structure of their internal organs in which the pongids
and hominids show quite remarkable resemblances, but for these details
information should be sought in more technical books. Similar parallels
are to be found in the physiology of the brain, the composition of the tis-
sues through which the young are nourished before birth, and even the
kinds of parasitic infestation to which man and anthropoid apes are susceptible.

In recent years the genetic affinities between these two families have
been affirmed still more dramatically. It has been known for years that
the higher apes and man have the same series of blood groups—depen-
dent, it is inferred, on identical genetic factors—and that they show very
closely similar immunity reactions of the blood as indicated by the pre-
cipitin reaction. In fact, serological reactions of the blood make it clear
that man has a far closer relationship with the anthropoid apes than with

Man's place in the
phylogeny of the primates
as reflected in
serum proteins
Class. III. Ev. '64

"The chromosome
of the Hominoidea"
?+ H E '64

the monkeys, and a recent careful study by Dr. Morris Goodman [31] of
the serum proteins has further indicated a particularly close relationship
between man and the African apes (gorilla and chimpanzee). Similar conclusions have been reached by Dr. Klinger [39] and his collaborators of the
Universities of Basel and London in their studies of chromosomes, that is,
the group of filaments in the cell nucleus containing the hereditary elements called genes. The results of this investigation have demonstrated
that the morphology and the total pattern of the chromosomes in man bear
a most striking resemblance to those of gorillas and chimpanzees (especially the latter), and in this respect all these are very different from the
monkeys. I should note here that these various lines of approach to determine natural classifications of living organisms form part of the general
subject of *taxonomy,* that is, the field of study concerned with classifications as a reflection of evolutionary relationships.

Now, the many anatomical, physiological, and biochemical features
that the hominids and pongids possess in common, and that distinguish
them as a zoological group from any other group of mammals, reasonably
may be taken as evidence of consanguinity, that is, that the hominids and

---

[1] Bracketed numbers in the text refer to the list of selected references at the back
of this book.

pongids really are genetically related in the evolutionary sense. This conclusion is simply an expression of the biological principle, based on many lines of scientific inquiry, that a unity of design among groups of animals is the result of inheritance from common ancestors of similar traits. Nevertheless the evidence of these common features is indirect. One of the complicating factors is the phenomenon of parallel evolution, the evolutionary development of the same feature in two separate groups of animals independently. It is known that this phenomenon may occur, and clearly may lead to resemblances that are in a sense fortuitous, and thus are not indicative of close affinity in the ordinarily accepted sense. On the other hand, even resemblances based on parallel evolution may be interpreted as indicating some degree of relationship, in so far as they are an expression of a well-known evolutionary principle that the same results tend to appear independently in descendants of the same ancestors. These descendants presumably must be endowed with rather similar inherited potentialities for evolution and therefore they will tend to react in the same way under the influence of similar environmental conditions.

Direct evidence for the evolutionary relationship of man and apes can only be supplied by palaeontological studies, that is, by the discovery and examination of fossil remains of past ages. For, if on the basis of indirect evidence intermediate phases of development from a common ancestral stock are postulated, then in the course of excavations in geological deposits relics of these phases might be expected to come to light in the search for fossils of extinct types. When Darwin wrote *The Descent of Man* in 1871, there was no fossil evidence of any significance to support his thesis of human evolution, and many critics of that time argued that this lack of evidence constituted a most serious objection to the whole thesis. Jibes about "missing links" provided plenty of ammunition for satirists and cartoonists who were strongly prejudiced against any suggestion that man, so "fearfully and wonderfully made" and more akin to angels than brutes, could possibly be related to such ugly caricatures of himself as gorillas and chimpanzees in the sense that they all took their origin from a common ancestral stock millions of years ago. However, three years before Darwin's *Origin of Species* appeared, an extraordinarily primitive and apelike cranium had been found in a cave in the Neanderthal valley near Düsseldorf, Germany, and since then many other remains of this same extinct type of man have been found. So distinctive are the skeletal characters of Neanderthal man (as this type is colloquially called) that some authorities regard the Neanderthal as representative of a species distinct from *Homo sapiens,* to which they have given the technical name *Homo neanderthalensis.* T. H. Huxley, in his essay "Man's Place in Nature" in 1863, gave close attention to the original Neanderthal skull, and stated that "under whatever aspect we view the cranium . . . we meet ape-like characters stamping it as the most pithecoid of human crania yet discovered." But he did not think the Neanderthal represented anything like a "missing link," for he went on to say that "at most [the Neanderthal remains] demonstrate the existence of a man whose skull may be said to revert somewhat towards the pithecoid type," and he was led to conclude at that time that "the fossil remains of Man hitherto discovered do not seem to take us appreciably nearer that lower pithecoid form by the modification of which he has, probably, become what he is." In this diagnosis, Huxley showed remarkable prescience, because it has indeed become evident that what are now termed the "extreme Neanderthaloids" were the result of a divergent evolutionary retrogression simulating certain pithecoid characters, but derived from precursors who were much more like modern *Homo sapiens* [50].

Almost thirty years later, in 1891, there were discovered in Java the skullcap and thighbone of a much earlier and more primitive type of man, called by its discoverer, Eugene Dubois, *Pithecanthropus erectus,* in other words, the "erect ape-man." Several crania, jaws and teeth, and portions of thighbones of the same group subsequently were found, not only in Java, but also in China near Pekin. The term *Pithecanthropus* seemed to be particularly apt, for when Dubois found and exhibited the first skullcap of this extinct type, anatomists were in some disagreement about its real significance, whether ape or man (see Fig. 1). Because of its general shape and size—the flattened braincase and its small capacity, the absence of a forehead, the enormous brow ridges, and the marked constriction of the cranium behind the orbital region—some argued that it belonged to a giant type of extinct gibbon. Others thought that the volume of the braincase, though small, did bring it into the human category. Still others took the view that it was neither ape nor man, but something in between these two groups. Later discoveries brought to light a number of jaws, some of which were of massive size with large teeth, but the latter, in spite of their unusual dimensions, were typically human in their detailed morphology and their disposition. The thighbones, in contrast with the skull, do not appear to be distinguishable from those of modern man and undoubtedly indicate that the pithecanthropines had already become as fully adapted to an erect bipedal gait and posture as *Homo sa-*

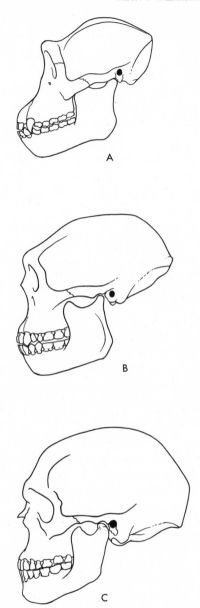

Figure 1. Outlines of the skulls of (A) a chimpanzee, (B) a pithecanthropine (*Homo erectus*), and (C) an Australian aboriginal. Note the intermediate position, anatomically, of the pithecanthropine. (Le Gros Clark, *The Antecedents of Man*, 2d ed., Edinburgh University Press.)

piens. Incidentally, this provides an example of "mosaic evolution," a well-established principle based on the observation that in the course of evolutionary change one part or system of the body may attain its final stage of development while other parts or systems still retain primitive characteristics; bodily structure does not always, or even usually, evolve as a whole in synchronous fashion.

The Javanese variety of *Pithecanthropus* was rather more primitive than the Chinese variety, with a smaller brain and more prominent jaws, and is reckoned to have lived about half a million years ago. The Chinese variety, with an antiquity of about 350,000 years, was more highly developed, with a larger brain and, in some cases, with quite a distinct, rounded forehead. Even so, there was a good deal of overlap between the two varieties in respect of their skeletal characters. For example, the cranial capacity of the Javanese skulls has been estimated to have ranged from 775 cc to at least 900 cc, and that of the Chinese skulls from 850 cc to 1300 cc. It is important to note that such a wide range of variation in the pithecanthropines is not limited to the cranial capacity; it is also quite apparent in skull shape and dimensions and in the dentition. However, a wide range of variation in morphological features is found in modern man and also in the different groups of modern anthropoid apes. This has been made abundantly clear in the very numerous publications of Professor A. H. Schultz of Zurich University [73, 74] as well as by other authorities. It is to be expected, therefore, that extinct groups of hominids and pongids would show at least an equivalent degree of variability. As a matter of fact, such variability is likely to have been even greater, for it is well established that during periods of rapid evolution (called by some explosive phases), when populations are expanding and spreading out into new territories and having to meet new environmental needs, individual variability may increase quite remarkably. This is very clearly exemplified in the initial evolution of the one-toed horses from three-toed horses. About ten million years ago, the first group of one-toed horses appeared in a genus called *Pliohippus.* According to Professor G. G. Simpson of Harvard University [83]: "Some of the primitive forms of *Pliohippus* still have minute side toes with tiny joints still projecting and some of the more advanced forms had evidently lost these. There was probably much variation in this respect, and even in a single group of *Pliohippus* some may have had the toes and some not." Furthermore, this variability in the single genus *Pliohippus* was by no means confined to the toes; it has also been noted in the teeth and skull.

It is of the utmost importance, in considering hominid evolution, to stress this sort of variability between individuals of the same group, and the even greater degree of variation to be expected between different geographical varieties of the same species or different species of the same genus. One gets the impression that some palaeontologists are ignorant of, or perhaps have just ignored, the literature recording such variability, for again and again they have been tempted to create new species or new genera on the basis of single or very fragmentary fossil remains simply because the latter are not *exactly* like other specimens that certainly belong to the same zoological group. It might almost be supposed that they expect individuals of the same species to be equivalent to identical twins! It is because of this misapprehension that discussions on the significance of *Pithecanthropus* have been much confused by the multiplicity of technical terms applied to the various remains. The Chinese fossils were originally, and indeed for some time, referred to a separate genus altogether, *Sinanthropus,* but this term has now been abandoned. A portion of a pithecanthropine skull with an upper jaw found in Java was called *Pithecanthropus robustus* as though it represented a completely new species, though there was no reason to suppose that it did not belong to a rather heavily built individual of the species already known as *Pithecanthropus erectus.* An unusually massive lower jaw found in Java in 1941 was actually referred to a different genus, *Meganthropus palaeojavanicus,* but, again, there is no good reason to think that this jaw was outside the range of variation of *Pithecanthropus erectus;* jaws of the single species *Homo sapiens* show just as wide a range of variation in size. Nor do these particular specimens exhaust the various scientific terms that have been conferred on different pithecanthropine remains, terms that have later been discarded because they have not been shown to be morphologically valid.

Probably nothing has done more to introduce confusion into the story of human evolution than the reckless propensity for inventing new (and sometimes unnecessarily complicated) names for fragmentary fossil relics that turn out eventually to belong to genera or species previously known. It has been very well said [36] that "rather than filling gaps which exist (in the fossil record), names tend to produce gaps that do not exist," and "it seems that some human palaeontologists regard the binomial system as a means of giving every hominid a Christian and surname and for creating phylogenetic schemes like family trees: an activity which starts with the nonsensical premise that the individual and not the population is the unit of evolutionary change." It should not be inferred,

G.A. Harrison + J.S. Weiner 1963 (4)? "Some considerations in the formulation of theories of human phylogeny" C + HE ed. SLWashburn of course, that the sequence of human evolution has been a simple matter of one species evolving into a succeeding species without any side-branches or collateral lines of development of which some no doubt became extinct. But the sequence certainly did not involve anything like the number of genera and species that have been invented from time to time by palaeontologists. We shall have occasion to refer to this subject again when we come to consider discoveries of primitive hominids a good deal older than the pithecanthropines. In this connection it cannot too strongly be emphasized that individuals do not form the units of evolution. *Populations* with a great genetic diversity comprise these units, and it is on their genetic diversity that the whole process of evolution actually depends.

I must now mention an important nomenclatural change arising from the recognition that the pithecanthropines constituted a very variable group; the generic term *Pithecanthropus,* by common consent of almost all anthropologists, recently has been abandoned altogether, the species now being referred to as *Homo erectus.* When the proposal for such a change was first put forward, I myself felt somewhat dubious about it, but I think this was largely because the generic term *Pithecanthropus* had been accepted for at least fifty years, and there is, perhaps, a natural reluctance to abolish suddenly so well-established a name. But on further consideration of the more advanced Chinese representatives of this ancient group, particularly the size of the larger brains (for the cranial capacity of some of them came well within the normal range of modern man) and the fact that the limb bones, so far as they are known, do not appear to differ from those of *Homo sapiens,* I came to the conclusion that it would be more in accord with the taxonomic principles of zoological classification to include them in the same genus, *Homo,* while still regarding them as specifically distinct from *Homo sapiens.* Apart from morphological considerations, there is good archaeological evidence that the Chinese pithecanthropines were well advanced in their cultural development; not only were they capable of fabricating stone implements with some skill, but the charred remains of their hearths make it clear that they had already at that early time learned to use fire for culinary purposes. Even though the technical term *Pithecanthropus* is no longer valid, we can still (partly to avoid misunderstandings that might be caused by such a sudden change of nomenclature) refer to the *Homo erectus* group by the colloquial name "pithecanthropines."

I have mentioned that the pithecanthropines (that is, *Homo erectus*) probably extended back in antiquity to about half a million years. It is convenient here to turn aside for a short statement on the methods available for estimating the antiquity of early types of fossil man. These methods can be divided into two categories, methods of relative dating and methods of absolute dating; for details of their practical application in palaeo-anthropology reference may be made to some of the systematic studies by Dr. Kenneth Oakley [61]. Relative dating methods are based primarily on the sequence in which geological deposits are laid down in successive phases of time. For example, sedimentary formations are the result of the deposition, layer by layer, of waterborne material such as sand or mud (which may become consolidated into hard sandstone or shale), or of stalagmitic material in caves, and it is clear that of the fossils which they may contain the older will be found in the deeper layers and the most recent in the uppermost layers. It is unusual in any one locality to find more than a few consecutive sedimentary layers out of the many that have been deposited over the ages, but by comparing one site with another it has usually been possible to arrange them in their proper time sequence. In tracing the story of human evolution these comparisons have been particularly important for the geological period called the Pleistocene which began almost three million years ago and during which some of the most interesting stages of the evolution of the Hominidae took place. In the latter part of the Pleistocene there were dramatic fluctuations of climate that were worldwide in distribution—from arctic to warm temperate, or (in regions of the earth that are today tropical) from very wet to very dry. These rhythmic climatic changes have been reflected in the composition and fossil contents of the geological deposits, and to some extent the latter can thus be correlated with each other stratigraphically and so provide a useful chronological scale of reference. In certain cases the relative antiquity of geological formations can be determined by the nature of the fossils they contain, for some species of extinct animals lasted through only a limited stretch of geological time and then died out. It is important to remember, however, evidence for the antiquity of fossil human remains by consideration of the associated faunal remains needs to be viewed with some care, for it does happen that a fossil species that has died out completely by the end of a certain geological period in one part of the world may have survived in another part to a much later time.

No doubt the most effective method of relative dating depends on the estimation of certain chemical elements that may accumulate in fossil

1964
FRAMEWORKS
FOR DATING
FOSSIL MAN

bones to an increasing extent over a long period of time. One of these elements is fluorine whose content in the bones increases with geological age. By a process of ionic interchange fluorine is very slowly taken up from percolating water in the soil and becomes fixed in bony tissue to form a very stable compound. However, although the amount thus increases with time, fluorine content also depends on the concentration of the element in the soil; for this reason the fluorine method does not permit a comparison of the relative antiquity of fossils found in different deposits in which the fluorine content of the soil may vary widely. But the method is invaluable for determining whether fossil human bones found in the same deposit in close association with the bones of extinct animals were contemporaneous with them, or whether, perhaps, they were the remains of a secondary interment. It was the fluorine test carried out by Dr. Oakley that first led him and his associates to suspect (and finally to prove) the fraudulent nature of the Piltdown skull and jaw. A similar method of relative dating has arisen from the observation that fossils may show some degree of radioactivity, the result of a gradual absorption of uranium where this is present; the degree of radioactivity thus depends in part on geological age.

Far more important than the relative methods of dating fossils are the absolute methods that have been developed and elaborated in recent years. One of these is the radiocarbon method. During life radioactive carbon is assimilated in a fixed amount into the organism from the atmosphere (directly or indirectly) and after death it undergoes a slow disintegration at a known rate. The proportion of radioactive carbon to its normal isotope thus progressively diminishes in dead organic material in relation to its antiquity. Lastly, mention must be made of the potassium-argon method. Natural potassium compounds in certain minerals, for example some of those found in volcanic ashes, contain a minute amount of a radioactive isotope of potassium which in the course of gradual decay disintegrates to form calcium and argon, and which has a half life of 1300 million years. Thus, again, by estimating the amount of radioactive potassium in relation to the calcium and argon present in appropriate minerals from a geological deposit, it is possible to calculate the antiquity of the latter (and therefore of its contained fossils) over a period of millions of years. It was the potassium-argon method that showed the Pleistocene period to have begun almost three million years ago (much earlier than had been supposed previously). The preceding geological period, the Pliocene, has been estimated by the same method to have commenced about twelve million years ago, and the period preceding the Pliocene, known as the Miocene, about twenty-five million years ago.

Now let us return to a consideration of the primitive hominids of half a million years ago, technically included in the species *Homo erectus,* but to which we may refer as the pithecanthropines. There is good evidence that populations of the same group were not limited to the Far East but spread their distribution into Europe and Africa. For example, there is the famous Heidelberg mandible found in 1908 in a stratum of sandy deposit that is assigned by most authorities on the basis of stratigraphic and faunal evidence to the first interglacial period of the Pleistocene, that is, somewhere about 400,000 years ago. In Algeria portions of the skull and jaws of a similar type have been given the name *Atlanthropus mauritanicus,* but almost certainly are referable to *Homo erectus;* the age of these remains is probably about 350,000 years. More recently in the remarkable stratified deposits in the Olduvai Gorge in Tanzania, Africa, Dr. Louis Leakey has unearthed a cranium very similar to that of the Far East pithecanthropines, except that the braincase appears to be a little larger than the mean value of the latter and the forehead rather better developed. The antiquity of this important specimen has been estimated by the potassium-argon method to be 490,000 years.

There is little reason to doubt that the species *Homo erectus* was ancestral to the species *Homo sapiens.* As indicated by fossil remains demonstrating the progressive increase in the size of the brain, the recession of the jaws, the development of a vertical forehead, and so forth, there is a graded morphological sequence linking the one with the other and this morphological sequence coincides quite neatly with a temporal sequence. In fact, the fossil record of the transition from the one species to the other could hardly be more complete. But if *Homo erectus* was ancestral to *Homo sapiens,* what are the characters to be expected in a still earlier species ancestral to *Homo erectus?* From the indirect evidence of comparative anatomy, and by analogy from what we know of the evolutionary history of other groups of mammals, it would be possible to make the following postulates. The mean cranial capacity of the pithecanthropines was about 1000 cc, a little more than two thirds of the mean cranial capacity of modern man; it may be inferred that in a still earlier ancestral stock the cranial capacity would have approximated more closely to the smaller capacity of the large apes. The pithecanthropines had large and prognathous jaws; their evolutionary predecessors probably had massive jaws of even more apelike dimensions. The teeth of the pithecanthropines were typically hominid in their general morphology, though rather large by modern standards; those of their predecessors would also presumably show the hominid pattern (and not the specialized features characteristic of the modern and most of the extinct genera of large apes), but they might be expected in some cases to be larger in conformity with larger and more powerful jaws. The limb bones of the pithecanthropines (so far as they are known) were quite similar to those of modern man, and, as we have seen, they appear to have reached the final stage of their evolution in the Hominidae long before the skull and jaw characters and the brain volume had done so. Such a seeming incongruity (an expression of mosaic evolution) suggests that earlier ancestral types would be likely also to show some degree of relative acceleration in the evolutionary advancement of definite hominid characters in the limb bones; although they almost certainly had already achieved the necessary anatomical basis for erect bipedalism, the functional perfection of this trait was a later culmination in the evolution of the genus *Homo*.

In summary, then, the immediate precursors of *Homo erectus*, hypothesized on the basis of indirect evidence drawn from comparative anatomy, comparative physiology, serological reactions, chromosome patterns, and so forth—evidence that has led to the assumption that the Hominidae originally developed from an ancestral stock in common with the anthropoid apes—must have been small-brained creatures with large, powerful jaws and teeth (the latter presumably hominid in character and lacking pongid specializations), and with limbs already adapted to some degree for an erect posture. If this line of reasoning is correct, then the remains of such an ancestral type might eventually be found in geological deposits predating those containing the fossilized remains of *Homo erectus,* perhaps with an antiquity of one or two million years. The remarkable thing is that this hypothesis has actually been verified by the discovery of just such primitive hominids, a group that is known under the generic name of *Australopithecus* or, colloquially, the australopithecines. It is with the story of their discovery that this book is concerned.

2

THE FIRST DISCOVERY
OF "AUSTRALOPITHECUS"

The story of the discovery of the australopithecines may be said to have begun with the appointment of Raymond Dart as Professor of Anatomy at the Witwatersrand University in Johannesburg at the end of 1922. I had known Dart before then, for in 1919 we were both demonstrators of anatomy in London University, he at University College and I at St. Thomas's Hospital. He was on the staff of the late Sir Grafton Elliot Smith, an anatomist of high reputation in his time, whose interests were primarily in the field of comparative neurology, and secondarily in problems of human evolution with particular reference to the evolutionary development of the human brain. Stimulated by Elliot Smith, Dart published a few papers on purely neurological matters before he left for South Africa.

I now turn to the actual discovery of the first specimen of *Australopithecus* and the reactions from other scientists to which this gave rise. The first formal report of this fossil to appear in print was an article in *Nature* on February 7, 1925 [16]. In this article Dart gave a preliminary description of the skull and an associated natural endocranial cast that he had

"A.a.: The Man-Ape of South Africa"

developed from a calcareous matrix in 1924.[1] The skull had been found in a fissure in a cliff of limestone formation at a place called Taung, near the Bechuanaland border in South Africa. There was no doubt that it was the skull of a hominoid type of some sort, either ape or possibly a very apelike hominid, and, as it was the first fossil of the kind that hitherto had been found in any part of Africa south of the Sahara, it was clearly an important discovery. In fact, Dart stated that his find was to be "logically regarded as a man-like ape." In addition he mentioned a number of features of the facial skeleton, the teeth, and the endocranial cast, in which the specimen seemed to him to make a closer approach to man than any of the known true apes do. He also drew attention to the palaeoclimatological evidence that the owner of the skull presumably lived, not in a tropical forest habitat for which modern apes have developed specialized adaptations, but in a relatively arid environment. He expressed the opinion that "the specimen is of importance because it exhibits an extinct race of apes *intermediate between living anthropoids and man,*" and gave it the name *Australopithecus africanus,* and he proposed "tentatively . . . that a new family of *Homo-simiadae* be created for the reception of the group of individuals which it represents." Now, it cannot be emphasized too strongly that Dart's article in *Nature* was a preliminary report embodying provisional conclusions, intended to be followed by a systematic monograph on the fossil specimen. But I think it was a pity that it was published in the form in which it appeared, because it included certain theoretical suppositions that were likely to raise doubts on the real significance of the skull, even though later these doubts were to be resolved in favor of Dart's interpretation of many of its anatomical features. At that time he had in South Africa no adequate comparative material of modern apes' skulls. Yet, from the relatively forward position of the foramen magnum on the base of the skull he suggested, with some caution it has to be admitted, that *Australopithecus* customarily assumed "an attitude appreciably more erect than that of modern anthropoids," but he had not been able to compare this with the range of variation of the position of the foramen magnum in modern apes. He then let his constructive imagination run on, in what some considered to be too rhetorical a style, to say:

> The improved poise of the head, and the better posture of the whole body framework which accompanied this alteration in the angle at which its

[1] An endocranial cast is a cast of the inside of the braincase. In its general form and proportions it provides valuable evidence of the general form and proportions of the brain itself, for the latter fits quite closely within the braincase.

Professor Raymond Dart, who reported on the first specimen of *Australopithecus*.

dominant member was supported, is of great significance. It means that a greater reliance was being placed by this group upon the feet as organs of progression, and that the hands were being freed from their more primitive function of accessory organs of locomotion. Bipedal animals, their hands were assuming a higher evolutionary role not only as delicate tactual, examining organs which were adding copiously to the animal's knowledge of its physical environment, but also as instruments of the growing intelligence in carrying out more elaborate, purposeful, and skilled movements, and as organs of offence and defence.

This style of writing was occasionally adopted by Elliot Smith in the popular expositions of his evolutionary speculations, but it was perhaps not appropriate for a scientific report in *Nature*. In fact, Dart's inference regarding the erect posture of the australopithecines was subsequently confirmed by further discoveries, but not solely for the reason that he advanced; the evidence of the position of the foramen magnum in the Taung skull, though it was suggestive, was not by itself adequate to draw such a conclusion. Dart's reasoning was more than lucky speculation; he made rather an inspired guess supported by certain other features of the skull that he had recognized but had not yet described in detail, such as the orientation of the orbital cavities (or eye sockets), the small size of the milk canine tooth, the slant of the forehead, and the relative size of the brain in so immature a creature (the milk dentition was still in place and the first permanent molar tooth had only just erupted).

I have mentioned that a natural endocranial cast composed of a sta-lagmitic matrix was found with the skull of *Australopithecus*. Now, Elliot Smith under whom Dart had studied in London had made a special study of the endocranial casts of fossil human skulls with the intention of finding out what might be learned from a delineation of the convolutions and fissures of the cerebral cortex that are to a limited extent imprinted on the inner surface of the cranial walls. Some of the inferences that he drew from these studies relating to intellectual development, manual dexterity, the acquisition of symbolic speech, and so forth, had drawn some rather strong criticism, partly because it was considered doubtful whether inferences of this sort could be made legitimately even from an examination of the brain itself, and also because it was questioned whether many of the elevations and depressions presenting themselves on endocranial casts of human skulls do really correspond to convolutions and fissures of the cerebral cortex as Elliot Smith had supposed. For example, after one of Elliot Smith's detailed accounts of the endocranial cast of a fossil skull, a paper was published in 1916 by Professor Symington of Belfast University [86] in which he compared endocranial casts of modern individuals examined in the postmortem room with their actual brains. The result of these comparisons was disconcerting; they showed that, in fact, very little of the convolutional pattern of the human brain can be defined with any accuracy on an endocranial cast, and, although Symington's general conclusions were contested by Elliot Smith, the whole matter of what one cynic has termed "palaeophrenology" was viewed with considerable scepticism. It was unfortunate, therefore, that in his preliminary article on *Australopithecus* Dart, in the initial enthusiasm aroused by his discovery, relied too much on insecure evidence of this type. He argued, among other things, that "the brain does not show that general pre- and post-Rolandic flattening characteristic of the living anthropoids, but presents a rounded and well-filled-out contour, which points to a symmetrical and

"Endocranial casts & Brain Form: Criticism of some recent speculations."
J. Anat. Physiol.

balanced development of the faculties of associative memory and intelligent activity." The introduction of dubious interpretations of this sort, following as it did the previous controversy on the whole subject of the kind of evidence regarding brain functions that can reasonably be expected from the study of endocranial casts, was certainly in part responsible for the critical attitude engendered by Dart's preliminary report.

Two other factors, I think, influenced this attitude. Before leaving University College, London, Dart had collaborated on a scientific paper putting forward a rather unorthodox view on the development of part of the nervous system, a view that many thought was based on inadequate evidence and which, indeed, not in accord with currently accepted views. Of course, there is no sound objection to the presentation of an unorthodox view simply because it is unorthodox; such presentations often lead to a most useful reappraisal of previous evidence and to fresh inquiries with new, more modern, and more efficient techniques. The paper to which I have referred, however, possibly may have led some anatomists to feel that Dart might be inclined too hastily to arrive at conclusions on too little evidence. The other factor was the extraordinary repetitious coincidence between Dart's discovery in South Africa and the discovery in Java by Dubois at the end of the last century. Dubois, who was a lecturer in anatomy at the University of Amsterdam, interested in comparative anatomy and human evolution, argued to himself that the likeliest place where relics of the postulated apelike precursors of modern man might be found was in the Indo-Malay region. In order to get opportunities to search for such fossil remains, he obtained a commission as a surgeon in the army of the Netherlands Indies and sailed for the Far East in 1887. In 1890 he actually found a jaw fragment, the first fossil evidence of the existence of an exceedingly primitive type of extinct man, which, as I have already mentioned, he later called *Pithecanthropus erectus*. In the following year he found the celebrated skullcap and femur belonging to the same hominid type, a type that, it is now generally agreed, was a representative of the ancestral stock from which *Homo sapiens* was derived. Dart, also a lecturer in anatomy with special interests in human evolution, sailed for South Africa in 1922, and by 1924 had in his laboratory an apelike skull which he regarded as "prehuman" in spite of its simian appearance, a "missing link" even more primitive than Dubois' *Pithecanthropus*.

The original discovery by Dubois was a surprising coincidence in itself—that in so short a time he should have discovered the very kind of fossil man for which he was looking. That there should be a second coincidence of an almost identical nature when Dart discovered what he also claimed to be a "missing link" seemed almost too much of a good thing; at any rate, combined with the few awkward features of Dart's preliminary article to which I have referred, it seems to have alerted the minds of anthropologists generally to the possibility that in his too enthusiastic zeal Dart had claimed far more for his australopithecine skull than was warranted by the evidence.

Reactions to the preliminary report in *Nature* came very rapidly in the next issue of this journal, when four leading anthropologists in England recorded their comments, and the very fact that *Australopithecus* was thought to be sufficiently important to receive such immediate attention in this way was significant in itself even though the comments on the whole were unfavorable to Dart's own interpretations. Sir Arthur Keith, who was then Conservator of the Hunterian Museum in the Royal College of Surgeons and recognized as a leading authority on fossil man and apes, concluded that "on the evidence now produced one is inclined to place *Australopithecus* in the same group or subfamily as the chimpanzee and gorilla," though he also admitted that it showed peculiar characters. Elliot Smith was cautious in his comments and remarked: "Many of the features cited by Professor Dart as evidence of human affinities . . . are not unknown in the young of giant anthropoids and even in the adult gibbon." He was impressed, however, by the evidence of the endocranial cast and wanted to know details of the exact form of the teeth. (It should be mentioned here that at the time of Dart's publication in *Nature* the teeth had not as yet been fully cleared of the stalagmitic matrix by which they were encrusted.) Dr. W. H. L. Duckworth, at that time an anatomist at Cambridge University, referred to a number of features of the skull that seemed to indicate a zoological status in advance of the large anthropoid apes, but concluded: "So far as the illustrations allow one to judge, the new form resembles the gorilla rather than the chimpanzee." The commentary by Smith Woodward, then in charge of the palaeontological collections at the British Museum, was a good deal more critical, and in reference to the question whether the direct ancestors of man are to be sought in Asia or Africa expressed the opinion that "the new fossil from South Africa certainly has little bearing on the question." Of these four referees, no doubt Smith Woodward was the least qualified to pronounce judgment on *Australopithecus*, for although he had written a few short reports on some palaeolithic human skulls and the jaw of a fossil ape, and had also reconstructed and described in detail the then famous (but now infamous) Piltdown skull, his authoritative field of work had been mainly confined to the study of fossil fishes and reptiles.

Other points of criticism were from time to time raised by students of fossil man. For example, Dart had laid some stress on the fact that the *Australopithecus* skull was long-headed, or, in craniological terms, dolichocephalic, and in this respect more hominid than pongid, and Keith had endorsed this opinion when he stated in his communication to *Nature:* "It is a true long-headed or dolichocephalic anthropoid—the first so far known." But in 1925 a paper appeared by Professor Bolk of Amsterdam in which he drew attention to the fact that dolichocephaly does sometimes occur among gorillas. Others carped about the name that Dart had conferred on his fossil skull, *Australopithecus,* calling it a "barbarous (Latin-Greek)" term. Actually, there is no technical objection to this term, though, as it has turned out, it has practical disadvantages. For one thing, it is now well recognized that the australopithecines are hominids and not apes in the proper sense; hence the suffix *"pithecus"* is rather inapt. For another thing, the term is not only unduly cumbrous to pronounce, but the prefix "Australo–" may give the casual and uninformed reader the impression that these creatures have some connection with Australia whereas the real meaning of the prefix is "southern." A much more pertinent criticism was leveled at Dart's tentative suggestion that a new family of the primates, *Homo-simiadae,* should be created for the reception of the group of individuals represented by *Australopithecus.* This, from the point of view of the taxonomist who is concerned with the natural classification of animals, was in fact an error of judgment. For "families" in the taxonomic sense are intended to represent quite different and contrasting lineages in the evolution of major groups of animals, and the dissimilarities between *Australopithecus* and either man or ape are far too slight to warrant a separate familial distinction. The whole issue at that time was whether the fossil specimen should be allocated to the ape family, Pongidae, or to the human family, Hominidae.

In 1931 Dart brought the skull of *Australopithecus* to England so that the original specimen could be inspected by anatomists and anthropologists, and he also distributed plaster casts to anatomists and palaeontologists in other countries. As a result, several authorities modified their opinions somewhat, but, apart from Professor Sollas who was then Professor of Geology at Oxford University, in the view of British palaeontologists and anatomists it was generally maintained that *Australopithecus* was an ape, though possessing a number of characters in which it showed some approximation to hominids, due, it was widely held, to parallel evolution that had taken place independently of the line of hominid evolution. Perhaps the most emphatic demonstration of the probable hominid affinities of the fossil was furnished by Dr. W. K. Gregory of the American Museum of Natural History [33] who, after a meticulous study of the dentition which had now been fully exposed by clearing away the calcareous incrustations that obscured its details, listed two characters in which it approximated more closely to the gorilla, one character nearer the chimpanzee and gorilla, and no less than twenty characters that he regarded as transitional to, or nearer to, primitive man.

I myself had seen the original skull in 1931, but I did not have the opportunity of studying it in detail, and I followed the general consensus held by my senior colleagues that *Australopithecus* should more properly be regarded as an ape. My first explicit reference to it in print was in an essay entitled "The Interpretation of Human Fossils" which appeared in a journal called *The Modern Quarterly* in 1939 [45]. There I wrote:

> This extinct ape had developed many features which were remarkably human (such as the characters of the dentition and the shape of the palate), so that some anatomists were led to suppose that it must bear an ancestral relationship to modern man. It is evident, however, that its geological age is too recent to permit acceptance of this hypothesis. Its main interest lies in the inference that, since it was almost certainly derived from the earlier *Dryopithecus* group [i.e., a very ancient group of generalized anthropoid apes that were in existence many millions of years ago], the latter was apparently endowed with the requisite potentialities for development in the direction of the Hominidae.

This argument, based on the estimated geological age of *Australopithecus,* was reasonable at the time, but it came to be invalidated later when it was demonstrated by further discoveries that the australopithecines date back to almost two million years. In a subsequent article [46], "Palaeontological Evidence Bearing on Human Evolution" in *Biological Reviews* (1940), I enlarged further on my discussion of the australopithecines, with particular reference to some further discoveries of the remains of these creatures made by the late Dr. Robert Broom of the Transvaal Museum, Pretoria, South Africa, and ascribed by him to different new genera of the australopithecines. Here I stated that, in spite of the fact that certain features of the skull and teeth show progressive trends approximating in some degree to a human level, "there is no doubt that these fossil genera are really apes and not primitive types of humanity."

Now, in all these early discussions we made the mistake of using the colloquial term "ape" without defining exactly what we meant by it. In some later controversies, also, the same loose usage of the term led to a good deal of confusion. Such confusion could have been avoided if we had used the scientific terms of zoological classification, Pongidae (instead of "apes") and Hominidae (instead of "man"). This distinction will be made apparent by reference to the accompanying diagram. (See Fig. 2.) We have already in the preceding chapter made brief reference to the evidence for a very close genetic relationship between the Hominidae and Pongidae, a relationship so close that these two families are commonly grouped together in generally accepted classifications in a single superfamily, Hominoidea. We have also noted that this implies that they had their evolutionary origin in a common stock that must have been essentially apelike in its primitive and generalized characters. In Figure 2 the anatomical characters of the common stock are schematically indicated by black circles. Some of these ancestral characters are of course inherited and retained in common by both families, and such characters may be termed *characters of common inheritance*. But when the two different lines of evolution start to segregate and diverge independently to form the separate groups of the Hominidae and Pongidae, each of these begins to acquire its own special and peculiar pattern of anatomical characters by which it comes to be distinguished from the other. Characters of this sort may be called *characters of independent acquisition*. In the diagram those that are distinctive of the Hominidae are indicated by crosses, and those of the Pongidae by white circles. Since the pongid sequence of evolution has been much more conservative than the progressive hominid sequence, its terminal products (the modern anthropoid apes) have preserved more of the original primitive characters of the common ancestral stock. As the divergent course of evolution proceeds, it will be appreciated that characters of common inheritance will become progressively supplemented or replaced by characters of independent acquisition in each line. Conversely, if the evolutionary lines are traced backward in retrospect they will be found to approximate more and more closely to each other in the characters of common inheritance which they share. Thus, for example, in representatives of an early stage in the hominid sequence (H^1 in the diagram) it will certainly be found that characters of common inheritance predominate over the characters of independent acquisition that are diagnostic of the Hominidae, the latter being as yet relatively few in number or only showing an incipient development.

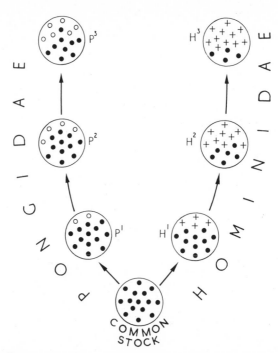

Figure 2. The diagram represents the divergence of two evolutionary sequences, the Pongidae (anthropoid ape family) and the Hominidae (the family that includes modern and extinct types of man). The two sequences inherited from a common ancestry *characters of common inheritance* (black circles). As the lines diverged each one progressively acquired its own distinctive features, or *characters of independent requisition;* those distinctive of the hominid sequence of evolution are represented by crosses and those of the pongid sequence by white circles. (Le Gros Clark, *The Antecedents of Man,* 2d ed., Edinburgh University Press.)

If, now, the remains of an individual corresponding to the stage H¹ is examined and its anatomical characters are compared quite indiscriminately—that is to say, if they are simply enumerated and summed without giving each one an appropriate weighting according to its evolutionary significance—the erroneous conclusion will be reached that, because in the *sum* of its characters it shows a closer resemblance to pongids than to modern man, therefore it should be classified as an ape. But this would be to ignore the highly important principle of "taxonomic relevance" in comparing anatomical characters as between the pongid and the hominid sequences of evolution. The decision as to the evolutionary status of a previously unknown fossil hominoid—whether it is a primitive representative of one or other of the two divergent lines of evolution corresponding to the two related families—must depend on a recognition of

the fundamentally different trends that have distinguished the evolutionary development of these two families and which are thus diagnostic of
each of them as a natural taxonomic group. The late Professor J. B. S.
Haldane [35] made this point when he said: "Palaeontologists rightly lay
more stress on differences which at a later date become the basis of familial or ordinal distinctions than on those which did not." In the particular
case of the Hominoidea the taxonomically relevant characters on which
the diagnosis of *pongid* or *hominid* depends are the characters of independent acquisition that serve to distinguish the divergent trends in the
two sequences, the nature and direction of these trends being at once
made evident by a consideration of the objectives actually reached by
their terminal products. The stage H[1] of hominid evolution is exemplified by the fossil genus *Australopithecus,* as we shall see. In the early
discussions and controversies following the discovery of this primitive
hominid, some anatomists, basing their judgment exclusively on the characters of common inheritance (such as the relatively small braincase and
the massive jaws) were erroneously led to suppose that it really was an
"ape" in the taxonomic sense. More careful studies soon made it clear
that many of the characters of independent acquisition distinctive of the
Hominidae, but none of those distinctive of the modern Pongidae, had
already begun their development and were superimposed on the characters of common inheritance. In other words, in those characters in which
Australopithecus had undergone modifications away from the common
ancestral stock of the Hominoidea, the direction of change was in the
hominid sequence and divergent from the pongid sequence.

From this brief discussion on the proper use of the scientific or taxonomic terms Pongidae and Hominidae, it will be realized how ridiculous
were the colloquial names "ape" and "man" when they were applied to
the question of the real nature of *Australopithecus*. Particularly was this
the case with the colloquial name "man," for some discussants were evidently using it as though it referred only to modern man, *Homo sapiens*.
But the name should be applied equally to prehistoric types such as
Neanderthal man and the pithecanthropines which, more especially the
latter, were characterized by a number of features that were astonishingly
"apelike." Of course, *Australopithecus* was very different in many ways
from *Homo sapiens* and it was never suggested otherwise. It did not differ
as much from the pithecanthropines (*Homo erectus*); indeed, on the basis
of the dental characters it is not always easy to draw a clear distinction in
some individual specimens between the two (though in this respect both
of them are quite sharply distinguished from the Pongidae). If we rely on

the distinctive characters of independent acquisition for determining whether a problematical fossil hominoid should be referred to an early phase in the hominid or the pongid sequence of evolution, there is some reason to suppose that the family Hominidae can be traced back even further than the australopithecine phase. The remains of an "apelike" creature, called *Ramapithecus,* have been found in Pliocene deposits in India and East Africa that have an antiquity of perhaps as much as five to ten million years, and this type, in its dentition and the shape of its palate, shows a clear approximation to the Hominidae and a corresponding contrast with the Pongidae. (See Chapter 5.)

As I have already noted, even Dart himself originally called *Australopithecus* an "ape," but he had the perceptive insight to recognize that it showed hominid features indicating that it was on, or at any rate close to, the line of evolution that eventually led to the final appearance of *Homo.* How correct his diagnosis was he could not realize at the time of his initial discovery, since it was only through the later discoveries made by Dr. Broom that the diagnosis became confirmed by the examination of numerous remains of skulls, jaws, teeth, and limb bones found at various other sites in South Africa. In his preliminary note in *Nature,* Dart said that he was preparing a detailed monograph on the skull found at Taung. This monograph never appeared, though he subsequently published a series of papers dealing with particulars of the dentition and skull [23], and in any case he was somewhat forestalled by the publication of an extensive monograph by a Viennese anatomist, Wolfgang Abel [1], based on the study of casts of the Taung specimen and entitled "Kritische Untersuchungen über *Australopithecus africanus* Dart." As a matter of fact, Dart had already completed his monograph and it was submitted to a journal for publication, but it was not accepted. I do not know why it was rejected, for I did not see the manuscript myself; possibly it was written in a style unsuitable for publication. I have sometimes wondered why some of the senior anatomists in London at the time (who fully recognized the obvious importance of Dart's fossil) did not advise him and help him to redraft his monograph in a form acceptable for publication. After all, Dart was only 32 years old at the time of his preliminary report on *Australopithecus,* he had made an outstanding discovery of the highest importance, and he may well have been excused in the excitement of his discovery for presenting a monograph written in a style that might be considered too effusive for a scientific publication traditionally accustomed to a more colorless presentation.

3

A VISIT
TO SOUTH AFRICA

In 1946 there appeared a monograph by Dr. Robert Broom of the Transvaal Museum entitled *The South African Fossil Ape-men: The Australopithecinae* [14]. In it he described in considerable detail the australopithecine remains that he himself had discovered during previous years at two sites in the Transvaal near Krugersdorp, Sterkfontein and Kromdraai. Short notes on the discoveries had also appeared as preliminary reports in *Nature* and other journals. Before recounting these I must say a few words about Broom himself, who was a very remarkable man and had earned high distinction for his palaeontological work. He was born in 1866, so that he was in his eightieth year when his monograph was published. His work during much of his lifetime had been concerned with fossil reptiles and their evolution, and particularly important were his studies of the mammal-like reptiles that flourished in South Africa over a hundred million years ago, some of which were undoubtedly ancestral to true mammals that later came into existence. Very soon after the first announcement had been made of the discovery of *Australopithecus* at Taung he went to Johannesburg to examine the skull and convinced himself that Dart's interpretation of this fossil was in general fundamentally correct. Indeed, for some years he was Dart's sole protagonist for the contention that in many of its anatomical features *Australopithecus* was properly to be regarded as a transitional stage between ape and man. He had, of course, visited the site at Taung, and he immediately set about looking for sites of similar geological formation in South Africa where further remains of this extinct creature might, perhaps, be found. He was rapidly successful in his search; in calcareous deposits at Sterkfontein and Kromdraai he found a remarkable series of australopithecine fossils including portions of skulls and jaws, many teeth, and some very informative limb-bone fragments.

The teeth, represented by several examples of the milk dentition and the permanent dentition, confirmed Dart's statement that they were essentially hominid in the details of their cusp pattern. The evidence of the skull fragments, as presented by Dr. Broom, was interesting, for although he emphasized the simian appearance of many of the cranial characters, he did point out some evident hominid characters, particularly in the mastoid region (behind the ear aperture), the formation of the glenoid cavity (that is, the joint cavity with which the condyle of the lower jaw articulates), the forward position of the occipital condyles on the base of the skull, and the construction of the skeletal elements of the cheek region. Of the limb-bone fragments, the lower end of the thighbone (femur) found at Sterfontein was evidently quite unlike that of any of the modern anthropoid apes and closely similar in many important details to that of modern and extinct types of man [8, 47]. Broom remarked in reference to this femoral fragment: "One thing is, I think, quite certain, the femur is that of an animal that walked, as does man, entirely or almost entirely on its hind feet." One of the bones of the wrist, the capitate bone, also found at Sterkfontein, was regarded by Broom as somewhat intermediate in form between that of a modern anthropoid ape and modern man. The lower end of an upper armbone (the humerus) and the upper end of one of the forearm bones (the ulna), both found at Kromdraai in close association with parts of australopithecine skulls, were even more impressive, because, judging from Broom's description and his illustrations, they were astonishingly like those of *Homo*. Another bone fragment found at Kromdraai (in the same block of stalagmitic matrix as the bones of the upper limb) proved to be an anklebone, called the talus, with which the shinbone articulates and through which the weight of the body is transmitted to the foot as a whole. Of this bone, Broom remarked that it "resembles that of man more than it does that any of the living anthropoids," and in his opinion it strengthened the supposition that the australopithecines were bipedal creatures. In a later chapter the significance of these limb bones will be discussed in some detail. (See Chapter 8.)

Dr. Broom further pointed out that the geological deposits in which the australopithecine remains were found indicated an arid or semiarid climate. Because he found associated with them in the same deposits at Sterkfontein the fossilized relics of an extinct hyaena which he identified as belonging to a genus *Lycyaena* that existed in Europe during Pliocene times, and also jaws of a sabre-tooth cat resembling a genus called *Meganthereon,* he provisionally suggested that the Sterkfontein deposits were of a late Pliocene date, and that the Taung deposit was even more ancient. This dating, it may be noted here, was questioned on the basis of subsequent studies. At any rate, he concluded: "These Primates agreed closely with man in many characters. They were almost certainly bipedal and they probably used their hands for the manipulation of implements. They certainly lived among the rocks and on the plains, and the Taung ape-man lived in desert conditions, while the Sterkfontein and Kromdraai ape-men lived in a land not unlike what the Transvaal is to-day." He further inferred, from the presence of horse remains at Kromdraai, that this particular deposit was of later date, Pleistocene.

In spite of all this additional, and apparently confirmatory, evidence that the australopithecines were apelike hominids rather than manlike pongids, and in spite of the fact that Broom did express his conclusions provisionally and with commendable restraint, his monograph still did not carry the conviction that might have been expected among anatomists and anthropologists generally. There were perhaps several reasons for this hesitancy. For one thing, the monograph, as it was written, gave the impression that it had been prepared rather hurriedly. Actually, this may have been so, for, as I have already mentioned, Dr. Broom was then in his eightieth year and as he was also suffering from periodical attacks of cardiac asthma he guessed he had not very long to live (in fact, he died five years later); he was naturally eager, therefore, to put on record his discoveries while he yet had time to do so. His illustrations, which were free-hand drawings by himself, also appeared rather crude and unfinished (though I came later to recognize that they were remarkably accurate). But the most unfortunate part of the monograph was the inclusion of a supplementary contribution by a young South African anatomist on the natural endocranial casts of the australopithecines that had been dis-

GWH Schepers
1946

covered [14]. Not only did this author persuade himself that he could identify almost all the convolutions and fissures on the lateral aspect of the cerebral hemispheres by a study of irregularities on the surface of the endocranial casts, he even delineated on the latter a great variety of different cortical areas which had been mapped out on the human cerebral cortex by microscopical studies and to which, on rather dubious evidence, specific functions had been ascribed from time to time by neurologists. He was thus led to such extravagances as, "The least we can say, therefore, is that these fossil types were capable . . . of interpreting their immediately visible, palpable and audible environment in such detail and with such discrimination that they had the subject matter for articulate speech well under control, and, having developed motoric centres for the appropriate application; they were also capable of communicating the acquired information to their families, friends and neighbours, thus establishing one of the first bonds of Man's complex social life." Based on such questionable evidence, this sort of precarious speculation was unfortunate to the extent that it naturally tended to discredit the whole monograph in the eyes of those who all along had been sceptical of the claims put forward for the hominid status of *Australopithecus*.

Apart from these controvertible suppositions about the brain, I should mention that, in spite of his recognized ability as a scientist, Dr. Broom had a peculiar propensity (not uncommon among palaeontologists working in the field and without the opportunity of acquainting themselves with the great range of variability in the anatomical features among single species, or even single subspecies, of modern primates) for creating new species or new genera on the basis of rather fragmentary material because of some relatively slight and trivial differences from type specimens previously described. Thus, Broom allocated the fossil material that he had found at Sterkfontein to a new genus which he called *Plesianthropus,* even though to most authorities (taking into account expected ranges of variability) it was really indistinguishable from the genus *Australopithecus*. Similarly, he regarded the australopithecine fossils from Kromdraai as representative of still another genus, *Paranthropus*. This unseemly multiplication of genera also no doubt contributed to the critical attitude taken by some anthropologists toward his monograph. As a matter of fact, the Kromdraai group of individuals was different in several respects from the Sterkfontein group and the *Australopithecus* specimen from Taung. This, indeed, might be expected from the geological evidence of a considerable difference in their antiquity; the Kromdraai deposits were estimated to date from the middle part of the Pleistocene period while the others are almost certainly early Pleistocene in date. The Kromdraai individuals were larger and of coarser build, and in some ways more "apelike" in their anatomical characters. For this reason they are now commonly assigned, not to a separate genus of australopithecines, but provisionally to a separate species, *Australopithecus robustus,* as compared with the Taung and Sterkfontein specimens which are included in the species *Australopithecus africanus.* But there are some authorities who are even doubtful of this specific distinction, and who prefer, therefore, to refer simply to the "robust" and "gracile" varieties of *Australopithecus.* Incidentally, there is here an interesting parallel with Neanderthal man of later Palaeolithic times, for the earlier neanderthaloids were more generalized and more like modern *Homo sapiens* than the extreme Neanderthals of later date who lived during the first climax of the last glaciation of the Ice Age. All the available evidence points to the conclusion that the extreme type of Neanderthal man was probably an aberrant type that became extinct, while the earlier generalized neanderthaloids probably gave rise to modern *Homo sapiens.* Similarly, it is probable that the earlier, more gracile, form of the australopithecines, that is, *Australopithecus africanus,* was ancestral to the genus *Homo,* and that the robust type represented an aberrant sideline of evolution that became extinct. At any rate, this is the view we take here, for it fits best with the geological and anatomical evidence.

Although I recognized that some parts of Broom's monograph were open to criticism, I was strongly impressed with most of the evidence that he presented regarding the hominid status of the australopithecines and I determined to examine the original specimens for myself if the opportunity should arise. The opportunity did arise when the first Pan-African Congress on Prehistory was organized by Dr. Louis Leakey to be convened in Nairobi, Kenya, at the beginning of 1947. I took this occasion to visit South Africa before going on to Nairobi to attend the Congress, but I had first made a careful examination and taken copious notes of the skulls and teeth of over a hundred modern anthropoid apes in various museum collections in order to assure myself of the normal variations that they show in their anatomical features and to compare them with the descriptions of the australopithecines already published by Dart and Broom. I still remained doubtful of the suggestion that the australopithecines were hominids rather than pongids, and I was still inclined to take the latter view. So far as the contentions of Dart and Broom were concerned, therefore, I felt I was going to South Africa as the "devil's advocate" in opposition to their claims.

I spent several days in Johannesburg, studying the Taung skull from every possible aspect, comparing it with pongid and hominid skulls in local collections, and making use of the notes I had accumulated from my previous studies of the much more extensive collections at home. I then went to Pretoria and spent the greater part of a fortnight at the Transvaal Museum examining all the specimens that Dr. Broom had collected at the Sterkfontein and Kromdraai sites. I also had the opportunity of visiting these sites to inspect the exact position where the fossils had been found. The results of my studies were very illuminating, not only because they made me realize how much more profitable it is to study original specimens rather than rely on plaster casts or photographs, but because they at last convinced me that Dart and Broom were essentially right in their assessment of the significance of the australopithecines as the probable precursors of more advanced types of the Hominidae. I gave an account of my observations at the meeting of the Pan-African Congress of Prehistory in Nairobi, and on my return to England I repeated these, in more detail, in a paper delivered before a meeting of the Anatomical Society in London. This paper [47] was subsequently published in the *Journal of Anatomy* in October 1947, under the title "Observations on the Anatomy of the Fossil Australopithecinae."

At the meeting, in the discussion following the paper I met with strong opposition from two members of the Society. One of these was the late Professor Wood Jones, a highly distinguished anatomist who, however, had long been an opponent of the commonly accepted thesis that hominids bore any close relationship to the pongids, and had developed the idea that "man" had evolved independently of monkeys and anthropoid apes from an ancestral group of small Eocene primates called tarsioids that had come into existence about sixty million years ago. He was therefore strongly averse to the suggestions that there could be any "missing links" such as the australopithecines which demonstrated so clearly a combination of primitive, apelike characters with advanced characters distinctive of hominids. My other opponent was Dr. (later Sir Solly) Zuckerman who had been a member of my staff at Oxford as a Demonstrator in Anatomy; he was sceptical of the construction I had put on the anatomical data, and also of that of Dart and Broom, and sought for statistical evidence of the "alleged" differences between the australopithecines and the modern anthropoid apes, in spite of the fact (which appeared obvious to me) that these differences, most of them severally and certainly all of them taken in combination with each other, were so sharp and so obtrusive on visual comparison as not to need statistical analysis in order to demonstrate that they were well substantiated. On rereading my paper of 1947, I find that my statements and conclusions, except for relatively minor points, were justified by the evidence then available. I drew attention to some significant hominid (and nonpongid) features of the braincase, the forehead, and cheek region of the skull, the shape of the jaws, the mixture of primitive characters (characters of common inheritance) and hominid characters in the capitate bone of the wrist and the talus bone of the ankle, the inference to be drawn from the lower end of the thighbone indicating an habitual upright posture, and, more especially, the essentially hominid characters of the dentition.

If I had to write the paper again, in the light of subsequent discoveries of australopithecine remains, I would modify my conclusions in a few respects. For example, I had inferred from the construction of the cheekbone in the robust type of *Australopithecus* that, as in the refined type, the eyebrow ridges were probably not strongly developed as they are in the modern African anthropoid apes. As it turned out later, in some individuals of the robust australopithecines these ridges were quite strongly built, even though showing certain differences from those of the gorilla and chimpanzee of equivalent size. I think, also, I laid too much emphasis on the hominid appearance of the lower end of the humerus and the upper end of the ulna, for in more lightly built chimpanzees it may be difficult (as pointed out by Professor W. L. Straus in 1948) to distinguish these particular skeletal elements even from those of *Homo* [85]. On the other hand, I had criticized G. W. H. Scheper's interpretation of the natural endocranial casts, and concluded: "From the general size and proportions of the endocranial casts it might be suggested that the mental powers of the australopithecines were probably not superior to those of the chimpanzee and gorilla, but this is the limit of legitimate speculation on the basis of such crude evidence. However, there is indirect evidence of quite a different nature which certainly makes it probable that the Australopithecinae were equipped with an intelligence definitely in advance of that possessed by the modern large apes," and here I referred to a quantity of baboon skulls excavated at the australopithecine sites, most of which showed signs of depressed fractures in the cranial vault suggesting that the injuries had been caused by a blow of some weapon used in hunting and killing these animals.

After the appearance of my paper certain criticisms of some of my conclusions were published by fellow anatomists. For example, a paper appeared in the *American Journal of Physical Anthropology* in 1949 by H. M. Kern and Dr. W. L. Straus on the lower end of the femur found at Sterkfontein [38]. These authors pointed out that in certain isolated metrical features the femoral fragment not only resembled the femur of modern man, but also that of some of the cercopithecoid monkeys. But they had not examined the original specimen, nor had they any casts of it for their study. In any case, it seemed to me that comparisons with the quadrupedal monkeys were beside the point, for, whatever else we might know about the australopithecines, we knew they were not quadrupedal monkeys; the real point at issue was whether the femur approximated more closely to known pongids than to known hominids. In fact, Kern and Straus did agree quite emphatically "that the fragmentary femur of *Plesianthropus* [that is, the type which is now known to belong to the genus *Australopithecus*] in general resembles the corresponding part of that bone in man and differs markedly from those of the anthropoid apes, is not to be contested." Unfortunately, at the time of their publication, the authors were not aware of the discovery of other fragments of limb bones and of the pelvis that made it clear that the australopithecines were adapted to an erect posture closely approximating to that characteristic of modern man though differing in some functional respects from the latter.

A more severe criticism was made about the dentition of *Australopithecus* and appeared in a letter published in *Nature* in 1950 in which it was affirmed that comparisons "made by appropriate statistical methods" led to conclusions that disagreed "with the claims of a number of workers [that is, Broom, Dart and myself] that in their size and shape the fossil teeth are in general more hominid than anthropoid" [94]. I questioned the truth of this statement in a subsequent letter to *Nature* with special reference to the milk canine teeth, to which a reply was made in a further communication that contrary to my expressed opinion "the milk canines of *Australopithecus africanus* and *Paranthropus* [*Australopithecus*] *robustus* do not differ in shape and dimensions, relative and absolute, from the chimpanzee." In the same year a paper appeared in the *Philosophical Transactions of the Royal Society* embodying a summary of the statistical data on which my critics had been relying for their own beliefs. It then appeared that the reference to shape and dimensions was simply based on the proportions of the overall length, breadth, and height of some of the teeth, and of others on only the length-breadth ratio. But obviously the "shape and dimensions" of a tooth can only be determined quantitatively by taking much more numerous and more discriminating measurements. It was manifest, also, that there must have been something radically wrong with the conclusions based on these statistics, for it was clear to anyone who had studied the original australopithecine teeth, and who were also well acquainted with the equivalent teeth in modern anthropoid apes, that in many respects they were very different indeed. Unfortunately, these metrical data were published in more than one scientific journal at about the same time; they therefore received a good deal of publicity, and no doubt many who had no specialized or personal knowledge of the subject may have been persuaded that conclusions based on statistical evidence (particularly since they were published in so reputable a journal as the *Philosophical Transactions of the Royal Society*) were final. I had myself, therefore, to carry out some further comparative studies, and also statistical studies, using more diagnostic measurements than those on which my critics had relied. For this purpose I examined the permanent dentition of 238 gorillas, 276 chimpanzees, and 39 orangutans, and the milk dentition of 89 gorillas, 105 chimpanzees, and 29 orangutans. The results of this study were completed in 1951 and appeared in the *Journal of the Royal Anthropological Institute* in the following year [48]. At about the same time, a letter appeared in *Nature* by Dr. J. Bronowski, scientific adviser to the National Coal Board, and his colleague Dr. W. M. Long, entitled "Statistical Methods in Anthropology"; in this they drew attention to the fallacy of treating anatomical characters statistically as though they were separate and independent characters instead of treating them in combination. They pointed out that a bone or a tooth is a unit and not a discrete assembly of independent measurements, and that to consider their measurements singly is likely to be both inconclusive and misleading. The right statistical method, they emphasized, must treat the set of variates as a single coherent matrix, and this can be done by the technique of multivariate analysis which is essentially a method (not possible with more elementary techniques) that can be used for comparing the *patterns* of anatomical structures. They applied this method of multivariate analysis to the australopithecine milk canines, demonstrating very positively their hominid character [4].

This controversy might have been prolonged indefinitely if Dr. F. Yates, a statistical expert, and his colleague Dr. M. Healy, puzzled (as they well might be) by the extraordinary discrepancy in the conclusions drawn from statistical analyses, had not sought to examine all the metrical data on which my critics had relied [93]. It was then that they discovered that, in making the relevant calculations, a vital step in arithmetic had been omitted in the final computation of standard deviations. When the appropriate correction was made, it became evident that the conclusions based on these calculations were to that extent invalidated. No doubt such an error can be attributed to the sort of chance oversight that may unwittingly befall any of us in the course of our investigations, but for a time it was rather a troublesome factor in our attempts to elucidate the true nature of the South African fossils.

To some it may seem a pity that this needless controversy regarding the australopithecine teeth should ever have occurred. In fact, however, it proved to be useful in several ways. For one thing, the error combined with its subsequent exposure served to demonstrate far more emphatically than before that the dentition of the australopithecines conformed to the hominid pattern and was very different from the pongid pattern. It demonstrated the inadequacy of elementary statistical methods for the determination of systematic affinities. It alerted the minds of biologists to the disquieting possibility that similar arithmetical errors of the same kind might not be so very uncommon in other biometrical papers. And it brought to the attention of biologists in general the value of multivariate analysis in comparing one zoological type with another. On the other hand, the outcome of the controversy did raise some doubts regarding the validity of statistical procedures for determining taxonomic relationships in defining natural classifications of animals. It was a pity that this should have been so, because it led some authorities to depreciate (perhaps too emphatically) the value of statistics in taxonomy. For example, in his Silliman Lectures of 1962 on "Mankind Evolving," the distinguished geneticist, Professor Dobzhansky [28], observed: "Mathematical statisticians (Pearson, Fisher, Mahalanobis and others) have devised techniques, called coefficients of racial likeness, discriminant functions, and generalized distance, which should decrease the art component and increase the scientific component in classification. These techniques are hopeful, but up to the present they have accomplished about as much for classification as the mechanical piano has for music."

There may be some who will demur at such a downright disparagement, but, in fact, there is a good deal of truth in Dobzhansky's remarks. For it needs to be recognized that while statistical comparisons may be of considerable value in determining differences between geographical varieties and subspecies, or even between species, there are serious difficulties in applying them to the differentiation of larger groups such as genera and families. In regard to the smaller groups they can determine consistent differences in, for example, the dimensions of the skull or the teeth, or, in the case of the races of mankind, the different distributions of hereditary traits such as the various blood groups, pigmentation of the skin, and hair texture. In larger taxonomic groups, the anatomical features may be too disparate to allow comparison of features that are sufficiently equivalent, a difficulty that involves what I have termed the "principle of morphological equivalence." For example, comparisons of the cranial height in different human races may be quite valid because the measurements are closely comparable and it gives an index of the height of the braincase at the level at which the measurements are made. But in comparing *Homo sapiens* with, for example, a large male gorilla it would clearly be misleading to employ the same craniometric technique, for in these animals the height of the skull is usually considerably extended by the development of a high bony crest on top of the cranium. If such a comparison were made, it would be a comparison of the height of the braincase in *Homo sapiens* with the height of the braincase *plus* the bony crest in the gorilla and would have no meaning from the taxonomic viewpoint. This is, of course, an extreme example, but it is not fully realized that similar (if less obvious) fallacies may be incurred in other craniometric comparisons in which overall measurements of the skull are commonly equated with one another. If the overall length of a modern European skull is compared with that of one of the pithecanthropine skulls quite serious difficulties are involved. In the European skull the maximal length is an approximate measurement of the maximal length of the braincase. But in the pithecanthropine skull it measures a good deal more, for the skull length in these primitive hominids involves a number of different elements that may be independently variable among themselves such as the exaggerated development of massive, projecting eyebrow ridges (forming the supraorbital torus), the greater thickness of the skull wall, and the projection backward of a prominent bony crest at the back of the cranium (the occipital torus). Many other examples of this lack of morphological equivalence could be given.

Another principle, the "principle of taxonomic relevance," may be mentioned here. In seeking to determine the true affinities of a fossil hominoid, some characters may be much more relevant taxonomically than others. For example, the limb bones of the pithecanthropines were very similar to those of *Homo sapiens,* while the skull, jaws, and teeth showed marked differences. Clearly, therefore, if the question arises whether the remains of a fossil hominid are those of one of the extinct pithecanthropines or of *Homo sapiens,* the anatomical features of the skull, jaws, and teeth are the relevant characters to which attention should be primarily directed. The point has been well put by Dr. Gwyn Thomas [87] in his warning: "It must be appreciated, however, that while fossil populations may differ statistically, the taxonomic significance of such differences must be a separate evaluation."

It has been necessary to enlarge on this matter of statistical methods as applied to taxonomy simply because it was the inadequacy of the methods which had been relied upon, and the arithmetical error in calculations of standard deviations, that did for some time lead to confusion in the minds of those not well acquainted with the australopithecine material, and, indeed, may still do so if reference should be made by anyone to those published statements on the dentition of these fossil hominids without being aware that they were inadvertently based on erroneous premises.

It is interesting in retrospect to note the rapid change of opinion regarding the true significance of the Australopithecinae that manifested itself when Dart's original communication to *Nature* was followed by one discovery after another of their fossilized remains. The most prompt and generous tribute to Dart's appraisal of their significance came from the late Sir Arthur Keith, a tribute the more notable because in his time he was certainly one of the foremost authorities on fossil man and human evolution. In a letter to *Nature* [37], Keith wrote: "When Professor Dart of the University of the Witwatersrand, Johannesburg, announced in *Nature* the discovery of a juvenile *Australopithecus* and claimed for it a human kinship, I was one of those who took the point of view that when the adult form was discovered it would prove to be nearer akin to the living African anthropoids—the gorilla and chimpanzee. . . . I am now convinced on the evidence submitted by Dr. Robert Broom that Professor Dart was right and I was wrong. The Australopithecinae are in or near the line which culminated in the human form."

My first visit to South Africa was of decisive value to me in my assessment of the true nature of the australopithecines. Since then I made two other visits there, in 1955 and again in 1964. On each of these occasions I took the opportunity to examine new material of *Australopithecus* that had been discovered since my first visit. I also visited the Makapansgat valley, about 200 miles north of Johannesburg where numerous remains of the genus have been found in one of the limestone caves that abound in the side of the valley, and where continuing excavations are likely to bring to light still further remains. I have here to acknowledge with gratitude the always ready help of Professor Dart and of Dr. (now Professor) J. T. Robinson (who succeeded the late Dr. Broom in his position as Curator of Fossil Vertebrates and Anthropology at the Transvaal Museum in Pretoria). Both these men, and also their associates, placed at my disposal all the fossil remains of the australopithecines that had been found in South Africa, and also personally conducted me to the various sites of their discovery. It is because of them that I can claim a rather intimate knowledge of all the factual data on which they themselves relied for stating their conclusions in their numerous reports on these fossil hominids.

ACCUMULATING DISCOVERIES
OF AUSTRALOPITHECINE
REMAINS

I have already noted in the previous chapter that, after Dart's initial announcement of the discovery of *Australopithecus* at Taung, many more finds of the remains of individuals of the same group were made elsewhere in the Transvaal by Dr. Robert Broom. I propose now to make further reference to these, and to anticipate more detailed discussion of them later by noting certain significant features which they display.

The specimen found at Taung comprised an incomplete skull but with the whole of the facial skeleton, and also a natural endocranial cast of a young individual—a juvenile with the milk dentition still in place—and it was therefore natural that there should be some speculation on the probable appearance of the skull of an adult individual and the degree to which it might differ from the skulls of adult anthropoid apes. I should mention, also, that it was found in the working face of a limestone quarry. The quarry was on the edge of a plateau of dolomitic limestone fissured by the solvent action of water percolating down from above and tunneled by underground springs. In the course of time, lime was deposited from the water in the form of stalactites and stalagmites, and as layers of a dense calcareous matrix called travertine. It was in such a matrix containing numerous bone fragments with fragments of rock (constituting what is called a bone breccia) that remains of extinct species of baboon and the first adult *Australopithecus* skull were found. The sites at Sterkfontein and Kromdraai, and also a further site at Swartkrans, were excavated by Broom between 1936 and 1949. These sites are situated quite close together not far from Pretoria and are also composed of bone breccias in caves or fissures. Relics of cave systems still exist at Sterkfontein, some of which have become completely filled in by lime deposits. Kromdraai and Swartkrans now appear on the surface as small kopjes, for there the cave roofs have been largely eroded away in the course of ages, leaving behind masses of bone breccias containing many more remains of the australopithecines mingled with skeletal fragments of a variety of extinct mammals. While Dr. Broom with his assistant J. T. Robinson were busy with these sites, Professor Dart was equally busy with the site at Makapansgat and here he assembled, year by year, an impressive collection of australopithecine remains of the greatest importance [17-25].

I have mentioned that Dr. Broom split up the australopithecine remains that he discovered into what later appeared to be an unwarranted number of different genera and species, but there is no need here to discuss them in detail. It is sufficient to note that the more gracile types of *Australopithecus,* that is *Australopithecus africanus,* were found at Taung and Sterkfontein, and that the more robust types, *Australopithecus robustus,* were found at Kromdraai and Swartkrans. But it is important to note that at Swartkrans were found two smaller jaws (one very incomplete) and a fragment of the front part of an upper jaw which were assigned to a separate and distinct genus, *Telanthropus.* It has been argued, mainly by Professor J. T. Robinson [64], that although, as he himself stated, this "type" shows australopithecine affinities, it is not an australopithecine in the strict sense, but represents a more advanced hominid that lived contemporaneously with *Australopithecus.* In fact, he later abandoned the generic name *Telanthropus* and allocated the specimens to the same species of early man previously known from Java, China, and elsewhere as *Homo erectus.* Robinson's conclusions were contested on the grounds that he had based them too insecurely on the small size of the teeth (as compared with the teeth of the robust australopithecines), on certain dimensions of the mandible, and on what some other authorities regarded as quite trivial and insignificant features of the front end of the upper jaw. In fact, in the gracile australopithecines the size of the teeth shows very considerable variation, and some found at Sterkfontein that are indubitably those of *Australopithecus africanus* compare quite well in size with those of the supposed *Telanthropus* or even with *Homo erectus*. Indeed, it needs to be recognized that, as I have already mentioned, it is not easy to distinguish some of the undoubted australopithecine teeth from those of *Homo erectus*. Dart [20] also demonstrated, by a careful analysis of the relevant data, that the dimensions of the mandible are really not adequate for separating those of *Telanthropus* from australopithecine mandibles that he had found in the limeworks cave at Makapan. Such differences as there might be, he noted, are no greater in degree than are found to occur in different individuals or racial varieties of the single species *Homo sapiens*. And the same could be said of the dentition. It needs to be recognized that the "diagnosis" of *Telanthropus* or *Homo erectus* as distinct from *Australopithecus* can hardly rest on the characters and size of the teeth and mandible, simply because in some individual specimens of either type these approximate very closely indeed, and may even overlap with one another. On the contrary, the differential diagnosis must rest on the quite characteristic and distinctive features of the skull, the cranial capacity, and the limb skeleton. In fact, nothing has been found in australopithecine deposits at Swartkrans, or elsewhere, to indicate with certainty that there existed contemporaneously with *Australopithecus* more advanced hominids with a cranial capacity and limb bones characteristic of the genus *Homo*. I have drawn attention in a previous chapter to the principle of taxonomic relevance in deciding the affinities of fragmentary fossil remains. Clearly, in the case of the australopithecines the teeth and jaws do not have a high taxonomic relevance for distinguishing between the genera *Australopithecus* and *Homo erectus*. It is the skull and cranial capacity, and the characters of the limb bones, that are the relevant characters for differentiating the two groups.

The cranial capacity of the South African australopithecines, as estimated from the very few skulls, or portions of skulls, available (in some cases partly broken or distorted), ranges from about 450 cc to something over 600 cc, that is to say, well within the range of variation of the largest of the modern anthropoid apes, the gorilla. But we do not know what the total population range may have been, particularly in the larger individuals in which some of the skulls have been severely distorted by crushing. Until quite recently the range of variation in the gorilla, based on several hundred specimens, was commonly stated to be from 340 cc to 685 cc, and then an out-size specimen was later reported to have a cranial capacity of 752 cc. The mean value of the *estimated* cranial capacity of *known* australopithecines has been computed to be 508 cc, and if the total range was anything like that of the gorilla (with a mean value of 498 cc), the upper limits must have been well above 600 cc. In fact, the cranial capacity of the immature Taung individual when adult has been estimated, on the assumption that the rate of brain growth was equivalent to that of the modern anthropoid apes, to have been perhaps as much as 620 cc. Professor Tobias [88] has attempted to assess the probable range for the Australopithecinae on the basis of the few (seven) known, and not all complete or undistorted, specimens, and arrived at the conclusion that "even the most generous estimate yields a maximum of 848 cc at the australopithecine grade of hominid evolution." Clearly, the evidence so far available is quite insufficient to arrive at anything like a firm conclusion on this matter.

It has already been mentioned briefly, in passing, that the limb bones of the australopithecines show that they were bipedal creatures, habitually adopting the erect posture that is distinctive of the Hominidae. But it needs to be well recognized, as several authorities have pointed out, that they had not achieved the fully erect posture and gait characteristic of *Homo sapiens.* For example, as we shall see, the femur, talus, and pelvis (particularly the latter) indicate that the transmission of weight from the trunk to the lower limb was different, and that the australopithecines were probably not capable of the unique "striding gait" of modern types of man (and presumably also of extinct types of *Homo* such as the pithecanthropines). However, the talus and the lower end of the femur are more closely similar to those of *Homo* than the pelvis, suggesting that the lower region of the leg and the foot had probably advanced further in adaptation towards a fully erect posture than the proximal parts of the lower extremity. This is what might naturally be expected, because until the foot and ankle region had undergone the modifications necessary for the achievement of a fully upright posture and gait, there could be no mechanical advantage in a modification of the pelvis and upper end of the femur to match changes in the foot that had not yet occurred. In other words, the progressive differentiation of the lower extremity for a fully erect gait is likely to have proceeded in a distoproximal direction, that is to say from the foot and ankle upward.

Unfortunately very few hand bones of the South African australopithecines are known, but two of them show highly significant features. A wristbone from Sterkfontein predominantly resembles that of *Homo,* but it does present certain primitive features which are still preserved in the modern anthropoid apes [47]. The other bone, a metacarpal bone of the thumb, from Swartkrans, is very robust and strongly curved; it likewise shows a curious combination of advanced charaters approximating to *Homo* and primitive characters that are to be found in the large apes. For example, there is a feature of the articular surface at the base of the bone in which it resembles the gorilla. (See Chapter 8.)

Now let me turn to more recent, and extremely important, discoveries reported by Dr. Leakey in East Africa. These discoveries were made in the steep and well-stratified sides of the Olduvai gorge, a magnificent gorge that stretches for several miles across the Serengeti Plain in the northern part of Tanzania, not far from the crater of Ngorongoro. The gorge owes its origin to river action cutting through layers of sediments some of which are composed of volcanic lavas. It is about 300 feet deep and, on the basis of geological stratigraphy has been divided into five beds, Bed 1, the lowest and oldest, dating from the Early Pleistocene.

In 1959, Dr. and Mrs. Leakey found a skull of the highest importance in Bed 1, about 20 feet below the top of this formation [41]. It proved to be an australopithecine skull which Leakey named *Zinjanthropus,* but which was recognized later to be quite similar to the skulls of *Australopithecus robustus* found by Dr. Broom at Kromdraai and Swartkrans in South Africa and therefore belonging to that species. The importance of this fossil is due first to the degree of its completeness (though broken into many fragments that had to be pieced together, it showed no very obvious distortion), and, second, to the fact that it could be dated by the potassium-argon method of absolute dating. It proved to be well over a million years old, in fact possibly as much as one and three-quarter million years, that is, presumably a good deal older that the South African representatives of the same genus. The skull was found on a "living floor" on which were strewn primitive stone implements of a type called Oldowan, as well as waste flakes struck off in their manufacture; by their close association they suggest that *Australopithecus* actually was responsible for fabricating them. Also found scattered on the same living floor were quantities of fossilized broken and splintered bones of various animals, evidently the remains of the australopithecine diet. Lastly, at the same site and on the same living floor were found two leg bones, the tibia and fibula. An examination of these bones led to the conclusion that the bony adaptation to bipedalism was well advanced at the ankle, while the knee joint region was probably much less well adapted for this purpose (see Chapter 8). It will be noted that this coincides with the inference drawn from the South African australopithecines that the differentiation of the lower extremity for functional bipedalism proceeded as a sort of evolutionary gradient from the distal toward the proximal end of the limb.

I now come to some veritable "bones of contention," and here, in spite of the fact that it will involve some repetition, I must anticipate more detailed references to this material in a later chapter in order to make it clear that I include these fossils within the general group of australopithecines. In 1960 and subsequent years Leakey reported the discovery of more hominid bones at a site (also a living floor) about 300 yards from the site just mentioned (where *Zinjanthropus* was found) but apparently at a very slightly lower level (about a foot or so). Thus there is no reason to suppose that these bones were geologically more ancient, and it is probable, indeed, the they were the remains of individuals that lived practically contemporaneously with *Zinjanthropus*. The bones included, among other fragments, portions of some of the cranial bones, the greater part of two lower jaws, a number of teeth, an almost complete skeleton of the foot, and a quantity of hand bones. Mainly because the foot bones show such a close resemblance to those of *Homo,* and because the braincase appeared to be somewhat larger than that of the very few South African australopithecine skulls in which it has been possible to get even an approximate estimate, Dr. Leakey, with his collaborators Professor Tobias and Dr. Napier, was led to conclude that these fossils did not belong to *Australopithecus,* but to a very primitive species of the genus *Homo* to which the name *Homo habilis* was given [44]. I do not find it possible to concur with this diagnosis, because (according to preliminary reports) in all its skeletal characters *Homo habilis* appears to correspond much more closely with the australopithecine group of early hominids. The cranial capacity has been estimated by Tobias to have been about 680 cc. This estimate, however, is based on two bones of the cranial roof, the parietal bones, both of which are cracked, fragmented, and incomplete; and even if this is accepted as an accurate estimate it still comes well within the gorilla range, as well as within estimations for the probable range of the australopithecines, and certainly well below the cranial capacity of any known specimen of *Homo.*[1] It may be noted that the low-

[handwritten margin note: 1964 Nature "A new sp. of the g. Homo from O...]

---

[1] It has been stated in one article that *Homo habilis* "had a more capacious cranial vault than any australopithecine." Such a statement is surely misleading, for it cannot be stressed too strongly that what is really meant is that the estimate for *Homo*

est estimate for the cranial capacity of one of the Javanese pithecanthropines (an estimate made by the late Dr. Weidenreich) was 775 cc, but this
figure was only arrived at by the consideration of one specimen of a skullcap in which the frontal and basal regions of the skull had to be
artificially, and somewhat speculatively, reconstructed. Clearly, such evidence as this is open to question; and, again, even if the estimate of 775 cc
is accepted, it is still about 100 cc more than the provisional estimate for
Homo habilis. Tobias himself has, in fact, rightly stressed the need for
reassessment of the capacities of the several Javanese crania of *Homo erectus.*

The foot skeleton from Olduvai, though very like that of modern
man in many respects, shows indications that the transmission of weight
to the foot was somewhat different and that the unique striding gait of
Homo had not yet been achieved, inferences also consistent with those
drawn from the pelvis of the South African australopithecines. Indeed, it
has been specifically stated that the talus shows some unusual features in
which it resembles the talus previously found at Kromdraai. The finger
bones, in their robust and strongly curved character agree very well with
the strongly curved thumb bone (metacarpal) found with australopithecine remains at Swartkrans. Further, the Olduvai wristbone that articulates with the thumb shows certain gorilloid features that parallel the
similar gorilloid features of the Swartkrans metacarpal already mentioned
(though in both cases it is clear that the thumb was opposable for grasping purposes in a manner characteristic of the human thumb). The teeth,
in spite of their small size, do not appear to be outside the great range of
variation of size in *Australopithecus,* and in some of their morphological
details do show resemblances to this genus (particularly in the shape of the
canine tooth). Such trivial differences in the length-breadth ratio of the
premolar teeth as have been reported are no greater than the differences in
the teeth among individuals of one of the anthropoid ape species, and Dr.
J. T. Robinson (probably the best authority on the australopithecine
dentition) can find in the dentition of *Homo habilis* no distinctive feature
that justifies its separation from the genus *Australopithecus* [72].

We may conclude, therefore, that the skeletal features of these ancient fossil remains are not only not incompatible—they are positively
compatible—with *Australopithecus,* and that the so-called *Homo habilis* is

---

habilis has been conjectured to be slightly greater than the few australopithecine skulls
so far discovered.

more properly to be regarded as a geographical variant, or just possibly a distinct local species, of that genus, only differing from the South African representatives of this group in rather minor details that indicate a more generalized and less specialized stage of development. (In the same way the earlier Neanderthals represented a more generalized and less specialized stage of development of the extreme Neanderthals of later date.) In any event, it would be surprising if the East African australopithecines, separated so widely both geographically and temporally from the South African australopithecines, did not show *some* differences from the latter. It has been suggested that *Homo habilis* is not morphologically distinguishable from the so-called *Telanthropus* (later called *Homo erectus*); this has led to the speculation that, all along, *Australopithecus* lived contemporaneously in the same localities with a more advanced hominid, and that the latter preyed on and devoured the former. Indeed, in one monograph there was figured a fanciful picture of the supposed more advanced hominids with the corpse of an australopithecine that they are imagined to have slaughtered for food. It is difficult to suppose that australopithecines and more advanced hominids (presumed to be representatives of the genus *Homo*) existed side by side in the same environment one and three-quarter million years ago, and still continued to live side by side about half a million years ago (as some have supposed to be the case in South Africa). Surely, in the course of more than a million years, the australopithecines would have been altogether wiped out to extinction by the predation of much more advanced and skillful hominids if the latter really had occupied the same habitat over this prolonged period of time! Lastly, even if *Homo habilis* is taken to represent a genus distinct from *Australopithecus,* the differences that it shows from previously known specimens of *Homo* are much too great to permit its allocation to the same genus. Here it is important to make comparisons with generic distinctions in other mammalian orders. For example, the well-recognized genus of horses, *Pliohippus,* is far closer in its evolutionary development and anatomical structure to the modern *Equus* than *Homo habilis* is to the modern genus *Homo*. In fact, it would be stretching the definition of the genus *Homo* beyond all reasonable bounds to include within it the diminutive-brained *Homo habilis,* with all its primitive (and evidently australopithecine) traits in the dentition and the limb skeleton. Of course, the confines within which the definition of a "genus" should be contained are not determined by any strict rules of zoological nomenclature, but some sense of proportion is required in drawing up such a definition.

It has seemed right to divaricate on these nomenclatural problems, for without doubt a great deal of unnecessary confusion has been introduced into the story of the australopithecines by the failure to recognize their high variability, thereby arbitrarily creating new species and genera that cannot be properly validated. It is not in dispute that two main groups of *Australopithecus* existed in Africa in ancient times, and that they may have been representatives of different species, *Australopithecus africanus* (a gracile type) and *Australopithecus robustus* (a more heavily built type). As an approximate parallel one may refer to the pygmy chimpanzee and the more robust common type of chimpanzee that now live on either bank of the Congo, or the forest gorilla and the mountain gorilla that are nearby neighbors in Central Africa, or the remarkable variability in the skull of adult orangutans in which some are relatively smooth and rounded and others of rugged build with massive bony crests on the top and back of the skull. In short, the precise relationship between the robust australopithecines and the gracile type (which in my view includes the smaller-toothed Skerkfontein fossils, as well as *Telanthropus* and *Homo habilis*), still awaits solution. They may well have been living together sympatrically (and in amicable, or at least unhostile, association) in exactly the same regions where their remains have been found.

It needs to be stressed that, at the time of writing, only preliminary reports of the remains of the so-called *Homo habilis* have been published and it may well be some years before they can all be studied in full detail. Meanwhile, one of the proponents of this new name has sought to forestall criticism by writing that "only a minority will immediately take the line of agreement or disagreement [that is, with the baptism of *Homo habilis*]; the majority will reserve judgment until they themselves have had the opportunity of studying the material and soberly considering its implications." It is perhaps permissible to express criticism of the validity of a nomenclatural decision when the authors of the latter have themselves expressed their decision on these same preliminary studies, and more particularly when my criticism is based partly on an inspection of some of the original material and casts of other specimens that I myself have been privileged to see and handle. I am indeed grateful to those who have kindly given me the opportunity to do so.

The question now arises whether there is any evidence that the australopithecines extended their geographical range beyond South Africa and the Olduvai Gorge. They certainly existed in other parts of northern Tanzania, for their remains have been found at Garusi (a fragment of a mandible with the two premolar teeth), and at Peninj near Lake Natron about 50 miles northeast of Olduvai. The Peninj find consists of an almost complete mandible that is extremely well preserved; it is reported to date from the early part of the Middle Pleistocene period, that is, considerably later than the skull of the robust australopithecine found in Bed 1 at Olduvai (*Zinjanthropus*). Yet it is said to fit the upper jaw of the latter almost perfectly. If the evidence of the antiquity of these two fossils is valid, robust individuals of *Australopithecus* must have been living in Tanzania over an astonishingly long period of time as a fairly widespread population in this area. It may also be emphasized once again that these East African individuals are certainly not specifically distinguishable from the robust types whose remains have been found at Kromdraai and Swartkrans in South Africa and whose antiquity has been estimated at about 500,000 years.

A fragmentary skull was found near Lake Chad by Dr. Coppens and briefly described by him in 1962 [15] as possibly the remains of an australopithecine skull. It corresponds in antiquity with known australopithecines in East Africa and South Africa; that is to say, it apparently predates any unequivocal specimen of *Homo* (for instance, *Homo erectus*), and Coppens's diagnosis thus may well be correct, although the cranial remnants are actually too fragmentary to allow of a certain diagnosis. Cranial fragments and teeth found at Ubeidiya in Israel, also assigned to the Early Pleistocene, have likewise been regarded as australopithecine in type, and they were associated with crude and primitive stone implements similar to those found with australopithecine remains in East Africa [84]. But, again, the diagnosis remains dubious because of the paucity and poor preservation of the skeletal remains. Lastly, mention must be made of several small fragments of large jaws with large teeth from the Djetis beds of Java, probably dating from the early part of the Middle Pleistocene and originally referred to a new genus of the pithecanthropines, *Meganthropus*. Robinson demonstrated a remarkable similarity, amounting in some features to a practical identity, between the cusp pattern of the teeth of *Meganthropus* and that of *Australopithecus* [65]. But, as I have already pointed out, in some cases it is difficult, or impossible, to distinguish between the dentitions of *Australopithecus* and the pithecanthropines; in other words, as I have emphasized, the teeth of these two groups of hominids are not taxonomically relevant characters for making a differential diagnosis between them. What is needed to establish Robinson's thesis that australopithecines existed in Java in the Middle Pleistocene is the discovery there of cranial remains, and also those of the postcranial skeleton. It seems more reasonable to infer (as I have argued elsewhere) that *Meganthropus* is after all nothing more than a representative of larger and more heavily built individuals of *Homo erectus* (for, in fact, the relative difference in jaw and tooth size between *Meganthropus* and accepted specimens of *Homo erectus* is no greater than may be found among the various races of the single species *Homo sapiens*).

It has to be admitted that, in spite of claims to the contrary, based on very incomplete and all too fragmentary specimens, there is as yet no certain evidence that the australopithecines extended their geographical distribution beyond that part of Africa that now lies south of the Sahara. It is, of course, possible that they may have done so, but, taking into account presumptive geographical and climatic barriers at the time, it seems improbable that they could have reached so far as Southeast Asia. They almost certainly evolved from simian ancestors on the African continent and they were evidently passing through a phase of hominid evolution when they were particularly vulnerable to predators and other hazards of the environment. They had as yet to develop an erect posture as efficient as that of *Homo* for running and walking; they had no powerful canine teeth for attack or defense; their hands were clumsily constructed and, although there is reasonable evidence that they used and even fabricated tools, the latter were very primitive and crude in design; and it may be inferred that their small brains would hardly have allowed them to cope with unexpected contingencies apart from those with which they had familiarized themselves in a relatively restricted type of environment. (See Chapter 9.) Only later, following a progressive development towards the *Homo erectus* phase of hominid evolution, with a larger brain (corresponding to a cranial capacity of perhaps 1000 cc or thereabout), with a perfected upright stance characteristic of the genus *Homo,* and with the manipulative ability to make more effective and more varied weapons and tools, only then would the possibility arise for adventuring into novel territories and eventually spreading over far regions of the Old World. Of course, this is no more than surmise, but it seems a reasonable surmise on the basis of all the evidence at present available.

5

THE HOMINID CHARACTER
OF AUSTRALOPITHECINE
TEETH

In this and ensuing chapters I propose to deal in greater anatomical detail with the skeletal features of *Australopithecus*. It is appropriate for several reasons to give first consideration to the dentition. For one thing, in the process of fossilization, teeth, by virtue of the hard enamel and dentine of which the crowns are composed, tend to be preserved more frequently and in more perfect condition than the purely bony elements of the skeleton. As a result, in the study of mammalian evolution from fossil remains, much more is known about the details of dental evolution than those of the evolution of many of the bony structures of the skeleton. This applies as much to the evolutionary differentiation of the pongids and hominids from a common ancestral stock as it does to the evolutionary differentiation of other mammalian groups. Again, it was the characters of the dentition of the australopithecines that first demonstrated quite unequivocally to all those well acquainted with primate odontology that these extinct creatures from South Africa were on the hominid line of evolution, approximating closely to the genus *Homo,* and distinct

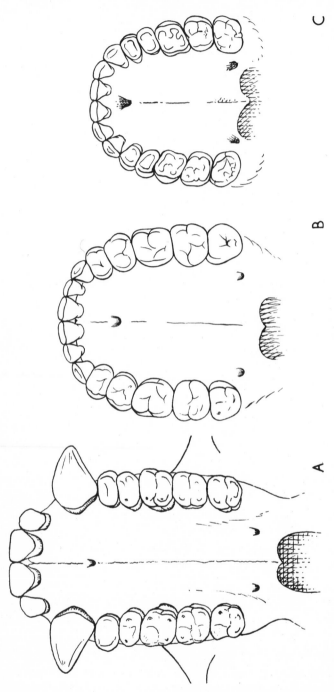

Figure 3. The palate and upper teeth of (A) a male gorilla, (B) *Australopithecus*, and (C) an Australian aboriginal. Note that in the curved contour of the dental arcade, the small canine teeth (worn down flat from the tip), and the absence of a gap (diastema) between the canine and incisor teeth, the total morphological pattern consistently presented by the australopithecine palate and dentition is fundamentally of the hominid type—and in sharp contrast to that of modern apes. (Le Gros Clark, *History of Primates*, 9th ed., by permission of the Trustees of the British Museum [Natural History].)

from the pongid line of evolution that culminated in the modern anthropoid apes.

Let us begin by comparing and constrasting the dentition of modern man with that of the modern large anthropoid apes (see Figs. 3 and 4). In many respects, of course, they are similar, particularly in the arrangement and pattern of the cusps of the molar teeth. In fact, as I have already mentioned, it may be very difficult to discriminate with certainty between single and isolated chimpanzee molar teeth and those of *Homo sapiens.* The same could be said about the molar teeth of some of the more generalized fossil apes of Miocene and Pliocene times. Pongids and hominids have the same dental formula; that is to say, they have on each side of the upper and lower jaw in the permanent dentition two incisor teeth, one canine tooth, two premolars, and three molars. The mutual similarities in dental anatomy have provided some of the most cogent evidence for linking together taxonomically the pongids and hominids as representatives of a common superfamily, the Hominoidea. But they are divergent representatives of this superfamily, and this divergence has been expressed by certain well-marked contrasts in the evolutionary development of their dentition.

In the modern large apes the incisor teeth (which like human teeth are spatulate in form) are much larger and broader than those of modern man. This hypertrophy of the incisors has been associated with a broadening of the front end of the jaws, and in the chin region of the mandible a thin ledge of bone called the simian shelf commonly bridges across the anterior end of the space between the two sides of the jaw. These are anatomical specializations that appeared in the later stages of pongid evolution, probably toward the end of the Pliocene, for in Miocene and early Pliocene fossil apes the incisors were smaller and, indeed, scarcely recognizable from modern human incisors, while the simian shelf was either absent or only incipiently developed.

In all large anthropoid apes, modern or extinct, the canine teeth are relatively large, conical in shape, tusklike, and sharply pointed (see Fig. 4). They show considerable variation in size among the different genera, and also between the sexes. The upper and lower canines overlap and interlock when the teeth are in occlusion with the mouth closed, the upper canine then coming into position behind the lower canine (see Fig. 9). There are several results of this interlocking. In the process of wear, at an early stage of attrition the canines show "attrition facets" on the anterior and posterior aspects of the crowns of the teeth where they become abrad-

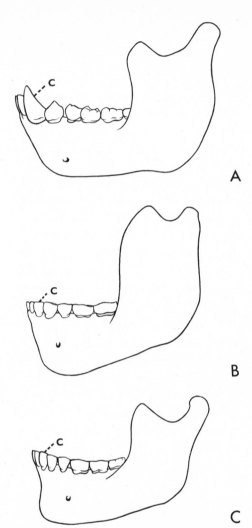

Figure 4. The mandible and lower dentition of (A) a male orangutan, (B) *Australopithecus,* and (C) *Homo sapiens.* Note the small size of the canine tooth in the australopithecine jaw, which becomes worn down flat from the tip to become level with the adjacent teeth in an early stage of attrition, as it does in hominids. (Le Gros Clark, *The Antecedents of Man,* 9th ed., Edinburgh University Press.)

ed by contact with each other and adjacent teeth. They do not wear down flat from the tip of the crown. Thus, except in very aged individuals where all the teeth show a very advanced state of extreme attrition, or where the canine crowns may have been accidentally broken off during life, the pointed canines remain projecting well beyond the occlusal level of the other teeth. In order to allow room for the interlocking canines in closure of the jaws a pronounced gap (called the diastema) almost invariably exists between the lateral incisor of the upper jaw and the upper canine. Lastly, it is obvious that the interlocking of the projecting canines is bound to modify and limit the rotatory movements of the jaws in chewing and thus affect the type of wear shown by the postcanine teeth.

In all known representatives of the genus *Homo* the canines are relatively small, spatulate in form, and bluntly pointed (see Figs. 3 and 4). On their eruption their pointed character may be rather obtrusive, but in the process of attrition they quickly become worn down flat from the tip and at an early stage of wear do not project beyond the occlusal level of the other teeth. Human canines have become assimilated into the incisor series of teeth from the functional point of view, since they are not used for specialized purposes of attack or defense as the fanglike canines of the large apes are used. It is interesting to note, however, that even in modern man the canine tooth is provided with a rather long and robust root, seemingly out of proportion to the modest functions that it is called upon to perform. This is one of the features that have been taken to indicate that the incisorlike canine of *Homo sapiens* has been derived in its evolutionary history from a more powerful type of tooth of simian dimensions and shape. The other features leading to this inference include the pointed character of the crown before it begins to be worn down (though the point serves no apparent function), and the late eruption of the tooth in the dental series. Further, in some fossil specimens of *Homo* (for example, *Homo erectus*) the canine may be larger than it normally is in *Homo sapiens*, and there may even be a small suggestion of a diastema between the upper canine and the lateral incisor. But no diastema of a size characteristic of the Pongidae has been found in any representative of the Hominidae, and, so far as is known, in all hominids whether recent or fossil the canines consistently wear down flat from the tip. Lastly in the Hominidae the canines show no pronounced sexual dimorphism as they do in the anthropoid apes.

The premolar teeth in hominids are bicuspid teeth. That is to say, the crown of the tooth is elevated into two cusps and these are set in an approximately transverse plane. In the anthropoid apes the premolar teeth also have two cusps of subequal size, except for the front lower premolar tooth. Now, this is an important tooth for the differential diagnosis of hominid and pongid dentitions. In the pongids it is what is called a sectorial tooth; the outer cusp is greatly enlarged and anteriorly forms a sloping surface that shears against the posterior margin of the upper canine in biting. The other cusp is much reduced, so that the tooth is predominantly unicuspid. Moreover, the tooth as a whole is commonly set obliquely to the axis of the line of the postcanine teeth. No doubt the sectorial character of the first lower premolar is a secondary adaptation which is related to the development of large, tusklike canines. But it is evidently a very ancient heritage in the pongid line of evolution, for it was already well developed in early Miocene apes.

The molar teeth of the anthropoid apes are distinguishable from those of modern man by their size (in the gorilla and the orangutan), by the fact that the fifth cusp at the back of the crown of the lower molars, the hypoconulid, is usually much better developed, and that, particularly in the last lower molar, the posterior half of the crown is commonly narrower than the anterior half. In the orangutan the enamel is elaborately crinkled, and to some extent this is also the case in the chimpanzee. In the gorilla the cusps are rather sharply angulated and they have been described as having a "crystalline" shape. The molar teeth of the chimpanzee are more similar to modern human teeth both in size and shape. Yet, although it may be difficult to distinguish some isolated teeth, particularly if they are severely worn so as to obscure the finer details of the cusp pattern, the latter is usually distinctive enough for their separate identification. In human teeth the molar cusps tend to be more rounded and closely compacted than in the Pongidae. Another characteristic feature of hominid molars is their mode of wear; at a very early stage of attrition the crowns become worn down to a relatively even, flat surface, even when the dentine underlying the enamel has as yet only very slightly been exposed. This mode of wear, which is combined with a similar type of wear in the canines and premolars so that at an early stage of attrition the occlusal aspect of the whole dental series forms a more or less even plane, is a hominid character that I have not found to occur in over 550 adult ape skulls that I have examined. Presumably it is related to a rotary motion of the jaws in chewing that (except in a modified form) is not possible in anthropoid apes because of their interlocking canines. Zuckerman [95] was led to the conclusion that the mode of wear of the molar teeth in the anthropoid apes is no different from that of human molars by demonstrating that the order in which the cusps become worn one after the other and the pattern of the dentine thus exposed are the same. Of course, this is perfectly true, but the crowns of the teeth do not become worn down in the *early* stages of attrition as they do in hominids; eventually they do, and in very old animals (particularly if the projecting canines have been broken off as sometimes occurs) the crowns of grossly worn molars may become quite flat. This mode of wear is very different from that characteristic of the Hominidae.

Another feature that distinguishes the pongid from the hominid dentition is the shape of the whole dental arcade. In the modern apes the canine and postcanine teeth form straight parallel rows that may even converge slightly toward the back (see Fig. 3). In the hominid dentition the teeth altogether form an even parabolic arcade and the canines, instead of being placed well behind the incisors as in pongids, are closely aligned with them to form part of a continuous and uninterrupted parabolic curve. The last molars are thus wider apart than the front molars and the premolars.

Equally conspicuous contrasts are to be seen in the milk, or deciduous, dentition. In pongids the canine is a sharp, pointed and projecting tooth, narrowing evenly from the base to the tip, while the first milk molar (the temporary precursor of the first permanent premolar) is a predominantly unicuspid and conical tooth of a sectorial character. In hominids the milk canine is spatulate in form, widening from the base to a little below the tip, while the first milk molar is multicuspid and not sectorial, the cusps (usually four in number) being approximately level with each other.

Now let us consider the australopithecine dentition. Fortunately there are available for a study a good number of complete, and almost complete, dentitions of *Australopithecus,* as well as several hundred isolated teeth. As regards the permanent dentition, the incisors are of typical human form; the canine teeth are spatulate and in the process of attrition become worn down flat from the tip with no attrition facets along the anterior and posterior margins; there is no diastema separating the upper canine from the lateral incisor tooth; the lower first premolar is a bicuspid tooth and not sectorial as in apes; the molar teeth wear down to a flat surface in the very early stages of attrition (see Fig. 7); and the dental arcade forms an even and uninterrupted parabolic curve (see Figs. 3 and 5). In the milk dentition the canines are spatulate as in modern man, and the first milk molar is a multicuspid tooth having cusps at approximately the same level on the surface of the crown (see Fig. 6).

There is good evidence from the study of immature specimens of the australopithecines that the order of eruption of their permanent teeth conforms to that of the Hominidae. It may be noted that in the ascending scale of the primates there is a progressive acceleration in the replacement of the deciduous dentition relative to the eruption of the permanent molar teeth, associated with a progressive prolongation of the growth period up to maturity of the individual. In the Pongidae the canine teeth

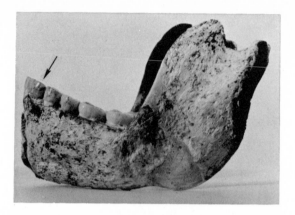

Figure 5. The mandible and lower teeth of *Australo-pithecus robustus* found at Swartkrans. Note the small size of the canine tooth, indicated by an arrow. (Courtesy of Dr. Robert Broom and Professor J. T. Robinson.)

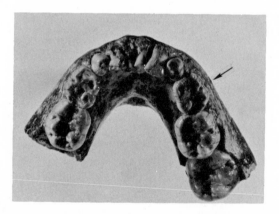

Figure 6. The jaw and teeth of an immature specimen of *Australopithecus robustus* found at Swartkrans, showing the milk dentition. Note the hominid character of the first milk molar tooth (indicated by an arrow) and the small size of the milk canine tooth immediately in front of it. The incisor teeth have been crushed back out of their normal position. (Courtesy of Broom and Robinson.)

are still very late in their replacement, often not completing their full eruption until after the last molar tooth has come into place, and in any case not until the second molar has completed its eruption. The late eruption of the canine was retained in certain fossil types of *Homo* (for example, *Homo erectus*), and is still retained occasionally even in some groups of *Homo sapiens*. In most modern races the canine erupts before the second molar, and the first incisor tooth may, in exceptional cases, actually erupt before the first permanent molar. According to Schultz [73], this occasional early eruption of the first incisor is a unique feature in which the Hominidae contrast with all other primates. From the fossil material collected by Dr. Broom at Swartkrans in South Africa, it is evident that in this particular group of the australopithecines the order of dental eruption corresponds with that normally found in *Homo sapiens* (that is, the canines erupted before the second molar), and in one specimen the first lower incisor tooth had erupted even before the first permanent molar. On the other hand, an immature mandible found at Makapansgat shows that in this particular individual the canine erupted after the second molar tooth (as in *Homo erectus* and among some modern races of mankind). The important point to emphasize, however, is that the pattern of dental replacement from the milk dentition to the permanent dentition that is characteristic of *Homo sapiens,* and also present in some of the australopithecines, has not been found to occur in any of the Pongidae, modern or extinct.

Finally, a point of some interest lies in the manner in which the molar teeth become successively and differentially worn down in the course of gradual attrition. It would be difficult to express such a feature quantitatively, but in the milk molars of the australopithecines the extent to which the dentine underlying the enamel has become exposed in some specimens is not, according to my own observations, paralleled by milk molars in anthropoid apes of a corresponding age. From the information at present available [73], in the chimpanzee the milk molars have completed their eruption by the seventeenth month and the first permanent molars come into position as early as the end of the third year. Thus there is a comparatively short interval of about nineteen months between the two periods of eruption when the milk molars are mostly exposed to attrition. In man the milk molars have usually erupted before the end of the second year, while the first permanent molar does not come into position till the sixth year. In this case, therefore, the interval between the two events is much longer (perhaps as much as four years). Thus the de-

Figure 7. The upper teeth (two premolars and the first two molars of an australopithecine jaw found at Sterkfontein, illustrating the flattened wear of the teeth in an early stage of attrition that is characteristic of hominids. (Courtesy of Broom and Robinson.)

gree of attrition of the milk molars in *Australopithecus* suggests that in the latter the eruption time of the first permanent molars may have been delayed as it is in man. If this inference is sound, it serves to confirm the implication of the delayed closure of the sutures of the cranial vault (see Chapter 6) that the growth period of the australopithecines was prolonged beyond that of the modern large apes. Such a conclusion is of considerable significance, for the prolongation of the growth period is highly characteristic of the human family. Attention may also be drawn to the relative degree of wear of the permanent teeth in the australopithecines. For example, in one specimen of *Australopithecus robustus* the third lower molar had evidently only recently come into position, for it is practically unworn (see Fig. 8). Yet the first molar is already planed down to an approximately flat surface with exposure of the dentine, while in the second molar only the minor grooves on the surface of the crown have disappeared and the main cusps are worn smooth. Such a degree of differential attrition in the three molars at this early stage of maturity is rarely, if ever, found in the anthropoid apes. Indeed, as pointed out by Schultz, noteworthy attrition of the teeth is unusual among the primates with an incompletely erupted permanent dentition. In man, however, it is not uncommon, partly, no doubt, in relation to the much longer interval between the eruptions of the molar teeth in the human dentition. According to Schultz's estimates, the interval between the eruption of the first and second molar, and the second and third molar, is in each case three and a half years in the chimpanzee and six and a half years in man. In other words, the length of time during which each molar is exposed to wear before the next comes into place is almost doubled in man. Thus, again, the degree of differential wear in the australopithecine permanent molars is strongly suggestive of an adolescent period similar to that of man, and considerably longer than that of the large apes of today.

It will be noted at once that in all its characteristic features, that is to say, in all those that are taxonomically relevant, the australopithecine dentition is fundamentally of the hominid type, and entirely divergent from the pongid type. Why, then, it may be asked, was there so much fruitless controversy on this matter when already the australopithecine dentition was quite well known from excavated material? Partly because, as I have already mentioned, inadequate and erroneous statistical results were published that purported to demonstrate no difference in "size or shape" between some of the fossil teeth and those of modern anthropoid apes. Partly also, perhaps, because some of the australopithecine teeth are

Figure 8. The lower three molar teeth from a jaw of *Australopithecus robustus* found at Swartkrans. In this particular specimen the first molar tooth had been flattened by considerable wear, while the second molar is only slightly worn, and the newly erupted last molar not at all.

of very large size, far larger than those of *Homo sapiens*. This is particularly the case with the robust variety of *Australopithecus*. But the size of a tooth by itself does not have a high valency for taxonomic distinctions, at any rate for the higher taxonomic categories. It may serve to distinguish subspecies or species, but not genera or families. And in any case we have noted that there is very considerable variability in the size of australopithecine teeth (just as there is in modern *Homo sapiens*), and some of them are no larger than the teeth of *Homo erectus*. Apart from these considerations, it was also the case that some of the early controversialists had never examined the original specimens for themselves, and some, indeed, were perhaps not as well acquainted as they might have been with the details of the comparative odontology of the higher primates. Be that as it may, it is now agreed by all those who can speak with authority that the dentition of *Australopithecus* conforms to the hominid pattern and not to the pongid pattern.

What can be learnt about the dietetic habits of the australopithecines from a study of their teeth? Not very much, I fear, and certainly not as much as some would suppose. It has been suggested that the robust types were predominantly vegetarian, partly because of the relatively small size of the canines and the large grinding surfaces of the molars, and partly because the enamel of the molars is chipped here and there in the same manner as those of creatures (like baboons) that subsist on coarse vegetable food containing gritty matter. On the other hand the gracile types, because of their somewhat larger canines and smaller molars are believed to have been more omnivorous and to a large extent flesh-eating. As a matter of fact, there is good evidence that both types of *Australopithecus* lived partly on a meat diet because animal bones, broken and split open for their marrow content, have been found in close association with their remains (see Chapter 9). This evidence is particularly striking at the living-floor site in Olduvai Gorge where Leakey found the massive skull of *Australopithecus robustus* to which he at first gave the name *Zinjanthropus*.

It has been argued that in some respects the teeth of *Australopithecus* were too specialized to have provided a basis for the subsequent evolution of teeth characteristic of the genus *Homo*. For example, in some specimens the molars are outsize in their major dimensions. But it is well attested from fossil evidence in other mammals that teeth may increase in size in an ancestral stock, only to decrease again in some of its descendants. Again, the canine tooth of *Australopithecus africanus* has rather a characteristic shape in the oblique and backwardly sloping anterior margin toward the tip of the tooth (incidentally, a characteristic of the so-called *Homo habilis* from Olduvai), and it may show an indentation of the enamel on the inner surface of the crown. The milk molar tooth, also, may be rather more elaborate in its cusp pattern than that of *Homo sapiens*. But these are trivial characters and their disappearance in subsequent evolution can be postulated without contravening any known genetic principles related to evolutionary change. After all, there is good evidence that the small canine of modern man has been derived by reduction from a much larger canine of simian dimensions and shape, and the first lower premolar tooth of *Homo sapiens* has undergone considerable modification by secondary reduction of the inner cusp which was much more prominent in the bicuspid tooth of ancestral forms. Thus there is no theoretical objection to the assumption that the australopithecine type of dentition (or, for that matter, the pithecanthropine type) could have given rise during the course of evolution to the type of dentition characteristic of modern man. It needs to be recognized, of course, that the dentition of modern man shows a high degree of variability among the different races of mankind in the size and morphological details of the teeth.

6

THE AUSTRALOPITHECINE
SKULL

Without doubt it was the shape and dimensions of the australopithecine skull that first inclined anatomists and anthropologists to the
view that *Australopithecus* was a true ape closely allied to the gorilla or
chimpanzee rather than a very primitive member of the family Hominidae. The small braincase combined with massive and projecting jaws certainly gave a superficial resemblance to the skull of anthropoid apes. But
as more skulls of young and fully adult australopithecines came to be discovered and their morphological details examined critically, this appearance was found to be indeed no more than superficial. It had not been
fully realized by some of those who at first refused to accept the conclusion of Dart, Broom, and others, that, in the early stages of hominid evolution when the Hominidae had (in geological terms) but recently become segregated from the Pongidae and had begun to embark as a separate line on their own evolutionary trend (divergent from the trend followed by the evolving Pongidae), the brain must have been relatively
small and the jaws large. It was of course already known that the pithecanthropines of half a million years ago had small brains—about two thirds
the size of the brains of modern man—and that the skull had a somewhat

apelike appearance by reason of the large and prognathous jaws (see Fig. 1). As I have argued in a previous chapter, it was therefore to be expected that in still earlier and more primitive representatives of the hominid line of evolution the brain would be even smaller and perhaps hardly exceeding simian dimensions, and that the jaws would similarly be more massive and projecting in a simian fashion. The australopithecine skull provides an example of just this stage in the evolutionary development of the hominid skull—more primitive than any other fossil hominid skulls previously discovered, but in a number of morphological details unmistakably a hominid skull and not a pongid skull. In retrospect it may seem surprising that students of fossil man were not ready to accept the obvious inference that the earlier, prepithecanthropine, stage of hominid evolution must have been characterized by small braincases and massive jaws, for it was generally held that hominids and pongids were derived from a common ancestral stock and such characters of common inheritance would certainly be shown in the initial stages of hominid evolution and perhaps persisted for some time. It was not till later, in the pithecanthropine stage, that the braincase began to expand toward a size characteristic of the genus *Homo* with the jaws undergoing a progressive retraction.

Let us now consider the main differences between the skull of a large anthropoid ape and a modern man. These differences are quite obtrusive, as may be seen by reference to Figure 9. The most conspicuous difference is to be seen in the size and shape of the braincase. In the ape it is much smaller, there is no rounded and approximately vertical forehead, nor is the back of the skull, the occipital region, rounded and smooth as in *Homo sapiens*. In the latter, also, the height of the skull above the level of the orbital cavities is very considerable. Viewed from behind, the sides of the braincase slope upward and inward toward the vertex in the ape so that the maximum diameter of the cranium is low down in the temporal region. In *Homo sapiens,* on the other hand, the maximum diameter is higher up in the parietal region. Combined with the large and powerful jaws in apes the small size of the braincase often does not provide a sufficiently extensive area for the attachment of the temporal muscles that are inserted below into the mandible. As a consequence, in adult male gorillas and orangutans, and occasionally in chimpanzees, these muscles reach right up to the midline of the cranium and as they grow to their full size build up a median crest on the top of the skull, the sagittal crest, to provide additional area for their attachment. Such a crest is never found in modern man, and the upper limit of the attachment of the tem-

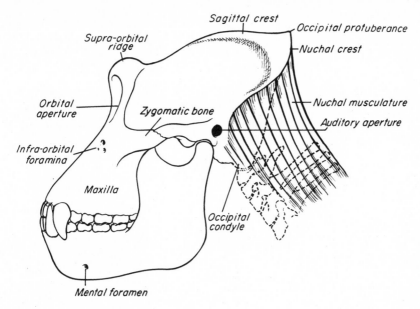

Figure 9. The diagrams illustrate the contrast between the pongid type of skull represented by a gorilla (above) and the hominid type represented by *Homo sapiens* (below). Note the massive development of the neck muscles in the large anthropoid ape, and the extensive area on the back of the skull for their attachment associated with a high nuchal crest.

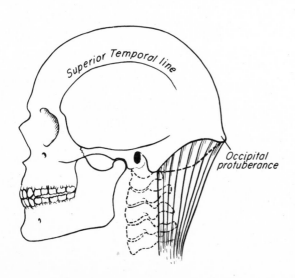

poral muscle is marked by a curved ridge (the superior temporal line) on the side of the skull. The height of this line, however, does show some variation in skulls of modern human races, and an instance has been recorded of an Eskimo skull (in which the jaws were unusually large and the temporal muscles unusually well developed) in which the temporal lines of the two sides almost reached each other at the top of the skull, but this is an exceptional case. In those of the large apes in which a sagittal crest is present it becomes accentuated posteriorly where it always meets the summit of a powerful nuchal crest at a pointed process called the occipital protuberance. The nuchal crest forms the boundary of an extensive, flattened area on the back of the skull that provides for the attachment of large and powerful neck muscles hafting the occipital aspect of the skull to the forwardly sloping neck vertebrae below. Such powerful muscles are necessary to hold up the skull with its heavy jaws in the oblique quadrupedal posture that the large modern apes normally assume on the ground, and this is particularly the case because the occipital condyles on the base of the skull, that is to say, the fulcrum about which up and down movements of the head occur at the joint with the upper end of the spinal column, are situated far back in relation to the whole length of the skull.

In *Homo sapiens* the condition is very different. Here the condyles are placed further forward and the jaws are much smaller and lighter. The skull is thus fairly nicely balanced on the top of a vertically disposed spinal column and there is therefore no need for a massive nuchal musculature to hold up the head in a horizontal position with the eyes looking directly forward; the occipital protuberance is situated at a low level (at the most not very much above the level of the auditory aperture), and the nuchal area for the attachment of the neck muscles is greatly reduced. In correlation with these differences, the long axis of the occipital condyle is approximately horizontal in *Homo,* whereas in the large apes it slopes upward and backward to some degree. It is important, also, to note the relations of the condyles to other structures at the base of the skull; for example, in *Homo* their midpoint is situated on a level with the posterior margin of the auditory aperture, while in apes (even in juvenile individuals) it is well behind this level.

The orbital apertures, that is, the opening of the orbital cavities or eye sockets, give the appearance of looking more directly forward in apes because their outer margins are more forwardly placed and not much of the orbital cavity is visible from the lateral aspect of the skull. In gorillas and chimpanzees the apertures are surmounted by a prominent ridge, the supra-orbital torus, that extends uninterruptedly across the midline. Moreover, they are placed well in advance of the front end of the brain-case. In *Homo sapiens,* on the other hand, the outer margins of the apertures are characteristically recessed and the orbits are placed below, rather than in front of, the frontal lobes of the brain. Below the orbital apertures is the infra-orbital foramen through which passes a sensory nerve, the maxillary division of the fifth cranial nerve; in the modern large apes this is usually a multiple foramen with two or three separate openings, but in *Homo* it is almost always a single opening.

The cheek region of the skull is in part formed by the zygomatic bone. This sends back a process, the temporal process, which articulates by a suture with a forwardly projecting process of the temporal bone, the zygomatic process, to form the zygomatic arch. In apes this arch is strongly built, for to it is attached the powerful masseter muscle which is inserted below into the outer surface of the ascending branch (or ramus) of the mandible. It should be noted, also, that the temporal process of the zygomatic bone is a long process, so that the junctional suture between it and the forwardly projecting process of the temporal bone is placed relatively far back. In *Homo sapiens* the zygomatic arch is slender and the temporal process of the zygomatic bone is quite short.

Behind the auditory aperture is the mastoid region of the skull. In the human skull a mastoid process is regularly found here, even in quite immature individuals; it is a pyramidal process of distinctive shape and provides for the attachment of some of the neck muscles such as the sternomastoid muscle that extends downward and forward along the side of the neck to reach the collarbone and the sternum. Although this process is commonly absent in the large apes it is occasionally to be found in adult male gorillas (but not in the skulls of young individuals). But even when it is present in old gorillas it differs from the human mastoid process in shape, having a flat posterior surface that is a lateral extension of the nuchal area of the occipital bone. Immediately in front of the auditory aperture is the bony socket. the mandibular or glenoid fossa, into which fits the condyle of the mandible to form the temporo-mandibular joint. In man the fossa is well hollowed out in a transverse direction and is bounded in front by a rounded eminence, the eminentia articularis, on to which the condyle of the mandible moves when the jaw is opened. Behind, the posterior wall of the fossa is formed by the bony wall of the auditory channel, the tympanic plate as it is called. In the modern apes the glenoid fossa is characteristically flattened so that the eminentia articularis does not stand out so conspicuously. The level of the fossa lies below the level of the auditory aperture and in this way contrasts with the human condition; moreover, its posterior wall is normally formed, not by the tympanic plate, but by a postglenoid process of the temporal bone that separates it from the plate. Lastly, in the ape the fossa is commonly bounded on the inside by a prominent entoglenoid process of the temporal bone. These differences between pongids and hominids in the constructional details of the glenoid fossa and its relations to neighboring structures reflect differences in the mechanism of the temporo-mandibular joint, and these again are reflected in the contrasting types of attritional wear of the teeth that I have already mentioned.

The massive upper and lower jaws of the large apes project forward in muzzlelike fashion, and into this marked prognathism the nasal apertures are recessed to an extent that the nasal skeleton is hardly visible from a side view of the skull. In *Homo sapiens* the retraction of the jaws has exposed a prominent nasal skeleton, and the sharp margins of the nasal aperture are prolonged in the midline below to form a projecting nasal spine. It may be noted that in primitive forms of the genus *Homo*, for example the pithecanthropines, in which the jaws were strongly prognathous as compared with modern man, the nasal skeleton is still relatively flat, and the nasal spine is not evident. Instead, as in apes, the floor of the nasal cavity is more or less evenly continuous with the front of the upper jaw into which it slopes along a low, rounded, bony gutter. In the hominids the bony palate forms an even, arcuate curve in conformity with the shape of the dental arcade already mentioned. In the modern apes, by contrast, the palate is elongated and of an approximately rectangular shape with parallel sides.

The massive lower jaw of the modern large apes differs from that of *Homo sapiens* in the absence of a projecting chin (instead, the anterior or symphysial surface of the jaw slopes downward and backward), in the shape of the articular condyle, and also in the large size of the ascending ramus that gives attachment to one of the masticatory muscles, the masseter muscle. It also differs in the position of the mental foramen through which passes a sensory nerve; this foramen is situated low down toward the lower margin of the mandible, while in *Homo* it is placed about halfway between the upper and lower margins of the jaw.

If they are viewed from the basal aspect, a number of differences between the ape and human skulls are very evident. The most obvious is the

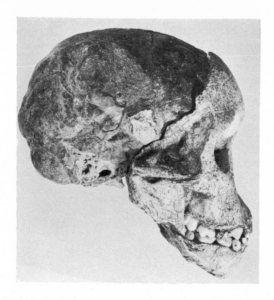

Figure 10. Side view of the immature skull of *Australopithecus africanus* found at Taung. Note the rounded forehead, the relatively short (orthognathous) facial skeleton, the recession of the outer margin of the orbital opening, and the small canine teeth. (Courtesy of Dart.)

position of the large opening, the foramen magnum, through which the brainstem passes to become continuous with the spinal cord. In the human skull it is situated far forward and its long axis is approximately horizontal. In apes it is placed farther back toward the posterior surface of the occipital region and its longitudinal axis is deflected upward and backward. These differences in the foramen magnum, like the difference in the position and slope of the occipital condyles to which I have already drawn attention, are, of course, related to the posture of the trunk and the poise of the head on the top of the vertebral column. Again, in the ape's skull the bony auditory canal is elongated and cylindrical in shape, while in the human skull it is relatively shorter and flattened anteriorly where it takes part in forming the posterior wall of the glenoid fossa. There are other differences in the disposition and size of the numerous vascular and nervous foramina in the base of the skull, but it is unnecessary to enter into details about these, except to note that the carotid foramen (which transmits the internal carotid artery to supply blood to the brain) is much smaller in the ape.

One further point about the skull needs attention, and that is the relative age at which the denticulate interlocking joints between the cranial bones, the sutures, become obliterated by osseous fusion between adjacent bones. In apes the cranial vault sutures begin to close very early, at about the time when the second permanent molar teeth erupt [40]. In modern man, the obliteration of the sutures begins much later, at about the thirtieth year, and proceeds more slowly. This difference is no doubt

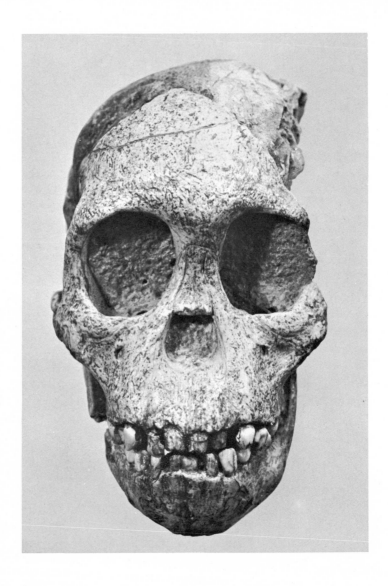

Figure 11. Front view of the immature skull of *Australopithecus africanus* found at Taung. (Courtesy of Dart.)

Dr. Robert Broom at the Sterkfontein australopithecine site. He is pointing to the position where he found part of an australopithecine skull.

related to the longer period of growth toward full maturity. There is some evidence that the time of fusion along the sutural lines is related to the growth of the brain. At any rate, in certain cases of microcephaly in man the sutures undergo premature closure, and deformities of the skull in young individuals can sometimes be corrected, at least partially, by the surgical procedure of opening up the bones on either side of the synostosed sutures.

Following this brief account of the structural contrasts between the skulls of the modern large apes and modern man, let us now give attention to the main features of the australopithecine skull. The first complete adult skull of the gracile type of *Australopithecus, Australopithecus africanus,* was found by Dr. Broom at Sterkfontein in 1947. It was almost perfectly preserved except for the lower jaw which was missing. Illustrations of this skull are shown in Figures 12 and 13, and in the latter the mandible has been reconstructed by reference to numerous other specimens. Incomplete skulls had been found at Sterkfontein in 1946, as well as several jaw fragments with the teeth *in situ.* The 1947 skull is that of a very aged individual, and all the teeth of the upper jaw had been lost. The most obvious feature of this skull is the combination of a small braincase with large jaws, and it is this that gives it such a simian appearance. It is well to emphasize once more these simian proportions of the australopithecine skull, because (as well recognized by Dart, Broom, and others) they are primitive characters in which the australopithecines contrast very strongly with *Homo sapiens* or even *Homo erectus.* But it is important to note that, while the general proportions are certainly "simian" in the sense that they approximate to the level of development still preserved by the modern anthropoid apes, they are not necessarily of taxonomic significance for the problem of deciding whether the australopithecines should be allocated in a natural classification to the Pongidae or to the Hominidae.

If the morphological details of the Sterkfontein skull are examined one by one in detail, and then in combination with each other, a surprising number of hominid (and nonpongid) features reveal themselves. The braincase is rounded and its vertex is elevated above the level of the upper margin of the orbital aperture far beyond that of the anthropoid apes, in this feature actually approaching close to *Homo sapiens.* This has been demonstrated statistically by what I have called the "supra-orbital height index." The occipital protuberance is low down as in hominids generally, so that the nuchal area for the attachment of the back neck

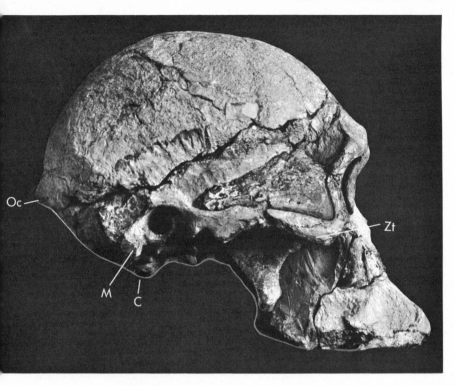

Figure 12. The australopithecine skull found at Sterkfontein in 1947. Note the low position of the occipital protuberance (OC), the well-developed mastoid process (M), the relatively forward position of the occipital condyle (C), and the forward position of the zygomatico-temporal suture (Zt). (Courtesy of Broom and Robinson.)

muscles is relatively limited as in modern man (this has been checked statistically by the "nuchal-area height index"). The occipital condyles are placed further forward than in the modern large apes in relation to the level of the auditory aperture and the total length of the skull (as confirmed by the "condylar-position index"), and their long axis is approximately horizontal instead of sloping upward and backward as they do in the large anthropoid apes. The facial part of the skull is more below the braincase than in front of it.

These hominid characters of the australopithecine skull are evidently correlated with a significant change in what is called the basicranial axis, that is, the axis formed by the longitudinal column of bones at the base

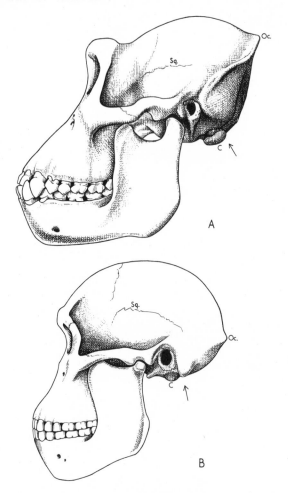

Figure 13. The skull of (A) a female gorilla compared with (B) the australo-pithecine skull found at Sterkfontein (*Australopithecus africanus*). In the latter the mandible and dentition have been reconstructed by reference to numerous other specimens. Note the relative positions of the occipital protuberance (Oc.), the occipital condyle (C), and the squamous suture (Sq.) between the temporal and parietal bones on the side of the skull. The arrow indicates the axis of the plane of the foramen magnum on the base of the skull. (Reprinted from *The Fossil Evidence for Human Evolution*, 2d ed., by Sir Wilfrid E. Le Gros Clark, © 1955 by the University of Chicago, by permission of the University of Chicago Press.)

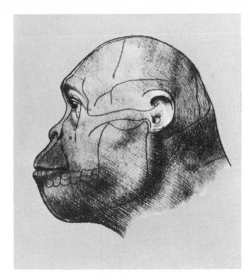

Figure 14. The diagram shows the outline of the almost complete australopithecine skull found at Sterkfontein with a reconstruction illustrating the probable appearance of the head during life superimposed upon it.

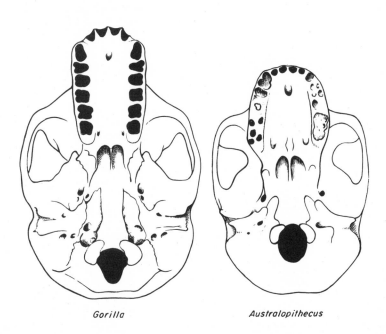

Gorilla   *Australopithecus*

Figure 15. A view of the base of the skull of a gorilla and that of the almost complete australopithecine skull found at Sterkfontein. In the latter the teeth were missing, and for better comparison, only the sockets of the gorilla teeth are shown. Note the contrasts in the shape and size of the palate, and the relative positions of the foramen magnum and the occipital condyles.

of the skull separating the braincase above from the facial skeleton below and in front. As may be seen by reference to Figure 16, in *Homo sapiens* the basicranial axis has two components. One of these is represented by a line drawn from the midpoint of the front margin of the foramen magnum, the basion (B), to a bony point at the anterior margin of the pituitary fossa, the presphenion (P), and the other by a line drawn from the point P to the midpoint of the suture between the nasal bones and the frontal bone, the nasion (N). These two components are bent on each other to form an angle opening downward. In the anthropoid ape skull, by contrast, they form almost a straight line (as shown in the median section of a gorilla skull in Fig. 16). The flexure of the basicranial axis in the genus *Homo* leads to significant changes in the skull. The downward bending of the anterior component brings the facial skeleton below rather than in front of the braincase and thus tends to diminish the prognathism of the jaws (even in those primitive hominid types in which these remain fairly massive). The relative bending forward of the posterior component brings forward the foramen magnum and the occipital condyles and at the same time alters their longitudinal axis from an oblique to a more horizontal position; this change in their position and orientation, in turn, allows the skull as a whole to become more perfectly balanced on the top of a vertical spinal column in association with an upright posture. Related to this, again, is a reduction of the nuchal area of the occipital bone and a lowering of the position of the occipital protuberance. Finally, the upward angulation of the two components leads to a deepening of the pituitary fossa and pushes up the braincase in relation to the facial skeleton so that the height of the braincase is raised even though the size of the latter is not markedly increased.

Attention should now be turned to the longitudinal section of the australopithecine skull found at Sterkfontein (Fig. 16). The outline of this section was prepared at my request by Dr. J. T. Robinson and is based on careful and accurate measurements taken by him of the original specimen. It will be observed that in this fossil skull there is a marked flexure of the basicranial axis in contrast with the condition characteristic of the large modern anthropoid apes, and associated with this are the several hominid features of the skull to which attention has already been drawn. In addition, the pituitary fossa is deep and sharply circumscribed as it is in *Homo*. It is to be noted that, although the flexure of the basicranial axis in the australopithecine skull is less acute than it is in modern man, the degree of angulation that it forms, accompanied as it is by

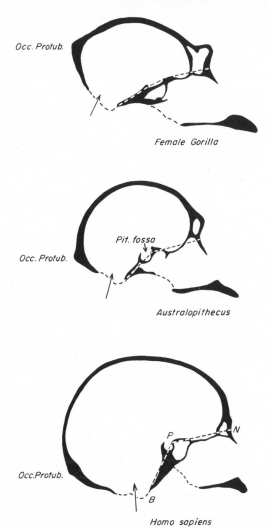

Figure 16. Longitudinal sections of the skulls of a gorilla, the Sterkfontein australopithecine skull, and modern man. In the australopithecine skull, note the hominid features of the components of the basicranial axis (BPN), the depth of the pituitary fossa, the low position of the occipital protuberance, the relative shortness of the face, and the axis of the plane of the foramen magnum (indicated by an arrow).

highly significant changes in cranial proportions and relationships, is a hominid character and contrasts strongly with the Pongidae. It was this complex of characters that first led to the provisional inference that the australopithecines had already achieved the erect bipedalism distinctive of the Hominidae, an inference that was later converted into a positive certainty by the discovery of the pelvic and lower limb skeleton.

Let me now mention a number of other, and more individual, features of the skull of *Australopithecus africanus* that are hominid in construction. Taken separately as isolated abstractions these features might not be accepted as of decisive importance for determining the taxonomic position of the australopithecines—whether they should be regarded as pongids or hominids. But, taken in combination with each other in constituting a "total morphological pattern," they do assume a significance of high validity. The orientation of the orbital aperture and the recession of its lateral margin, particularly as seen in the original Taung skull (Fig. 10), are typically hominid in character. The general orientation of the zygomatic bone contrasts rather strongly with that found in all recent large anthropoid apes; in the latter the facial surface of the bone looks not only laterally but also forward to a considerable degree, being shunted in this direction as the result of the lateral prominence of the heavily built zygomatic arch behind it. Further, this surface slopes forward gradually into the facial surface of the maxillary bone of the upper jaw without any abrupt change of inclination. In *Australopithecus* the facial surface of the zygomatic bone is orientated more as it is in the human skull, its main part being directed laterally so that it is angulated rather abruptly at the zygomatico-maxillary suture with the infra-orbital part of the maxilla which is here directed mainly forward. In *Australopithecus* also (as in other hominids), the upper border of the temporal process at its junction with the posterior border of the main part of the zygomatic bone is approximately at the same horizontal level as the lower margin of the orbital aperture. Further, the temporal process is abbreviated, and not prolonged backward as it is in apes.

In the maxilla the infra-orbital foramen is single as normally in hominid skulls, and the contour of the palate is rounded in conformity with the parabolic dental arcade. This contrast with the modern anthropoid apes is demonstrated very clearly in Figure 15, showing the basal aspect of the skull in a gorilla compared with the well-preserved Sterkfontein skull. Incidentally, this same illustration also emphasizes the relatively forward position of the foramen magnum and occipital condyles in

Australopithecus, and the consequent relative abbreviation and width of that part of the occipital bone that lies immediately in front of them, the basi-occipital bone.

The glenoid fossa of the temporo-mandibular joint in the australopithecine skull, which is well hollowed out, is bounded anteriorly by a transversely disposed and rounded eminentia articularis, thus reproducing very closely the human condition and very unlike the broad flattened surface in the gorilla or chimpanzee. In both these apes also (but particularly in the gorilla), the articular surface is prolonged downward along its inner margin into a strong and rather prominent lip formed by the entoglenoid process, but, as in human skulls, this is usually only slightly indicated in the australopithecines. Behind the glenoid fossa there projects down a bony tubercle from the posterior end of the zygomatic arch, the postglenoid process. This process is powerfully developed in the modern large apes and (as already mentioned) is usually interposed between the articular condyle of the mandible and the tympanic bone which forms the auditory canal. It varies in size in different australopithecine skulls, but in some specimens it is quite a small tubercle of remarkably human appearance. The tympanic bone is thus commonly well exposed below the process and presents a flattened surface that actually forms throughout a part, or (as in man) the whole, of its extent the posterior wall of the glenoid fossa. The bony auditory canal in the australopithecine skull is also relatively shorter than the more elongated and more tubular structure in the ape's skull.

Behind the auditory aperture, every australopithecine skull so far discovered, whether of young or mature individuals, shows a pyramidal mastoid process of typical human form and relationships. This is a highly significant difference from the modern anthropoid apes in which a mastoid process, if it develops at all, only develops after the attainment of full maturity, and even then is not entirely comparable with the human mastoid process in shape and relationships.

The australopithecine mandible is a massive structure in association with the large size of the grinding teeth (Fig. 5). There is no projecting chin, because the symphysial region slopes downward and backward as it does in anthropoid apes and also in primitive types of *Homo* such as the pithecanthropines. On the other hand, there is no simian shelf, and the mental foramen is displaced upward instead of near the lower margin of the jaw as it is in the Pongidae. The tooth-bearing part of the jaw is also more rounded and more abbreviated than in the large modern apes.

The cranial sutures are late in their closure, another contrast with the pongids. This significant feature is particularly well demonstrated in a skull found by Dart and his associates in a skull at Makapansgat (Fig. 17). In this interesting specimen the sutures on the vault and back of the cranium still remain clearly patent even though the third molar teeth had fully erupted and show a marked degree of attrition [24].

Reference must now be made to some of the skulls of the robust type of *Australopithecus (Australopithecus robustus)* found at Swartkrans in South Africa and at Olduvai in East Africa, in which the vertex of the cranium shows a low sagittal crest in the midline. When such a crest was first noted and described by Broom, it was assumed by some that it was strictly comparable with the sagittal crest normally present in adult male gorillas, and that it must have been associated, as it is in these animals, with a high and strongly developed nuchal crest (or occipital torus as it is sometimes termed). Thus, Zuckerman [95] wrote: "The question is, is it conceivable that it [i.e., the sagittal crest] was not associated with an occipital torus of the kind possessed by the gorilla, and with the powerful neck muscles of this ape? *A priori* the answer is no." He further argued that unless *Australopithecus robustus* "is the one exception to a morphogenetic process common to all known Primates, it follows that its possession of a high sagittal crest presupposes the presence of a powerful and shelf-like occipital torus." Apart from the fact that the australopithecine crest is not so very "high"—indeed, it is rather low in comparison with many gorillas—Zuckerman's line of reasoning demonstrates the dangers of arguing from the general to the particular, for, as it has turned out, *Australopithecus,* being a hominid, is indeed an exception to nonhominid primates in this particular feature. But it is also the case that, when Zuckerman's comments on the sagittal crest of the Swartkrans australopithecines were published (in 1954), already at that time, unknown to him, the occipital region of some skulls had been found at Swartkrans that showed clearly enough the absence of the powerful and shelflike torus that he had postulated. It was perfectly evident from these specimens that in the skull of *Australopithecus robustus* the occipital protuberance and nuchal crest were situated low down on the back of the cranium and that the nuchal area for the attachment of the neck muscles was limited in extent, as it is in hominids generally. In other words the sagittal crest in these australopithecines did not extend back to merge with a high and prominent nuchal crest as it does in the large apes. This was made particularly manifest in the almost complete australopithecine skull found by Dr. and Mrs.

Figure 17 (left). The back view of an australopithecine skull found by Professor Dart at Makapansgat. The cranial sutures are still widely open though the last molar tooth had already erupted and showed a marked degree of attrition. Note also the low position of the occipital protuberance and the absence of a high nuchal crest. (The scale is in centimeters.) (*American Journal of Physical Anthropology*, Vol. 20 (1962), p. 123.)

Figure 18 (below). The skull of *Australopithecus robustus* found at Swartkrans. Note the sagittal crest (S) on the top of the skull and the well-developed mastoid process (M). The back of the skull is missing, and the whole skull has been somewhat distorted by crushing. (Courtesy of Broom and Robinson.)

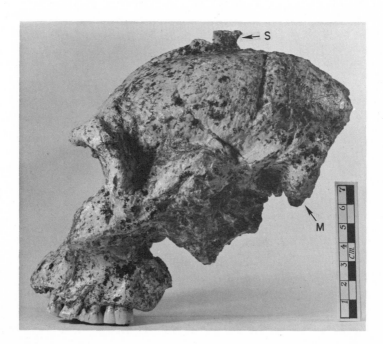

Leakey at Olduvai in Tanzania. In this remarkable specimen there is a low sagittal crest (as in some of the Swartkrans skulls), but it does not extend back into the occipital region to meet a "powerful and shelf-like occipital torus." On the contrary, the occipital protuberance is separated by a considerable distance from the posterior end of the crest and, as in all the australopithecine skulls found elsewhere, it is placed very low down with a corresponding restriction of the nuchal area as it is in more advanced hominids.

It needs to be emphasized that, apart from the development in mature individuals of a low sagittal crest, the skull of the robust type of *Australopithecus* shows quite pronounced contrasts with the gracile type. For example, the brow ridges are more heavily built, the vertex of the cranium is lower in relation to the level of the supra-orbital margin, the zygomatic arch and the jaws are more massive, and the grinding teeth are larger in size. The skull is also characterized, like that of the gracile type, by many of the same hominid features. The parallel that has been suggested with Neanderthal man and the modern type of man is certainly very apt. By some authorities these two types are regarded as separate species—*Homo neanderthalensis* and *Homo sapiens*—but by others they are both included in the species *Homo sapiens;* yet the morphological contrasts between them are no less conspicuous and pronounced than the contrasts between the skulls of *Australopithecus robustus* and *Australopithecus africanus,* and it is therefore a matter for discussion whether these two australopithecine types are to be regarded taxonomically as separate distinct species, or whether, after all, they are no more than two distinct subspecies or varieties of the same species.

The criticism may be made that some of the hominid features of the australopithecine skull that I have listed have not been subjected to statistical analysis and that to that extent the conclusions drawn from them have not been completely validated. This applies, for example, to the configuration of the temporo-mandibular joint, the orientation of the zygomatic bone, the form and relationships of the mastoid process, and the appearance of the orbital aperture. Such characters are sometimes referred to as qualitative rather than quantitative characters. In fact, however, no such distinction is really legitimate for all so-called qualitative morphological characters are ultimately capable of being expressed quantitatively. To do would require the application of an immense amount of highly complicated statistical procedures. So much so, indeed, that they are hardly worth the amount of work and time that this would involve. It

Figure 19. Dr. and Mrs. Leakey examining the upper jaw of a large autralo-pithecine skull they discovered at the Olduvai Gorge site in Tanzania. (By per-mission of the Armand Denis Productions.)

seems adequate, therefore, to reply on direct visual comparisons of these features in pongids, australopithecines, and the more advanced hominids, provided that such comparisons are the result of experience based on the study of sufficiently large numbers of skulls of the different types and that due attention is paid to the range of variation in each type. This I had made my aim by compiling careful notes of more than 100 ape skulls before comparing the latter with the original specimens of australopithecine skulls found in Africa that I have been able to examine for myself. It may be that some "qualitative" feature of the australopithecine skull that I have taken to be a typically hominid feature might be found by diligent search to occur as a rare variation among anthropoid apes. This would not invalidate such conclusions as I have drawn from it unless it can be shown to occur *consistently in a random sample* of anthropoid apes as it does consistently in the random sample of australopithecine skulls that have been discovered. I must also emphasize that the inference as to the essentially hominid nature of the australopithecine skull does not depend on any single character considered as an isolated abstraction, but on the whole pattern of characters considered in combination with each other and in combination with the various cranial indices that have been subjected to statistical analysis.

Finally, let me reiterate that the apparently "simian" characters of the australopithecine skull, such as the small braincase and the massive and projecting jaws, are characters of common inheritance from a hominoid ancestry that also gave rise by divergent evolution to the modern anthropoid apes and thus do not indicate a close taxonomic relationship with the latter. On the other hand, the hominid characters are characters of independent acquisition clearly demonstrating that the australopithecines were representatives of the hominid (and not the pongid) line of evolution. In other words, on the evidence of the skull alone they are to be grouped within the evolving trend of the Hominidae and taxonomically, therefore, are to be included in this family.

7

THE AUSTRALOPITHECINE
PELVIS

No part of the postcranial skeleton shows a more marked contrast between the modern anthropoid apes and modern man than the bony pelvis. In the former it is constructed on the same common pattern as that of
quadrupedal mammals in general. In *Homo sapiens* it has become pro-
foundly modified in adaptation to an erect posture and a bipedal gait.
Compare, for example, the outlines of the main element of the pelvis, the
os innominatum, in apes and man as shown in Figures 20 and 21. In apes
the blade of the os innominatum, the ilium, is elongated and narrow, and
its upper border, the crest of the ilium, is relatively short. The iliac bones
are also splayed wide apart so that their deep surfaces face almost directly
forward. In *Homo sapiens* the iliac bones are short and relatively broad,
surmounted by a stout crest elongated from before backward, and they
have become rotated forward so that their deep surfaces (forming the iliac
fossae) now face inward to a considerable degree, and the iliac crests converge forward. The crest of the ilium, viewed from above, shows a pronounced sigmoid curve. These changes in the pelvis of modern man are
all related to his characteristic erect bipedalism. The broad iliac blade
provides appropriate attachment for the powerful muscles of the buttock,

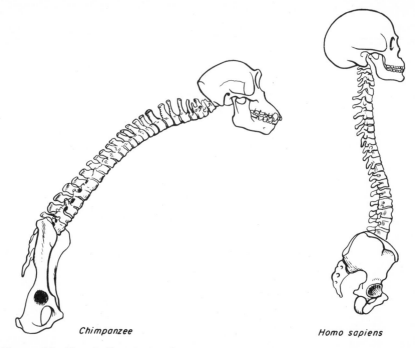

Chimpanzee   *Homo sapiens*

Figure 20. The skull, spinal column, and pelvis of a chimpanzee compared with those of *Homo sapiens*. Note the contrasts in form and orientation of the pelvis, and the well-marked curvatures of the spine in modern man.

the gluteal muscles, which are used for essential movements at the hip joint and for balancing the trunk on the thighbones in standing and walking. The rotation of the ilium alters the axis of pull of some of these muscles so that they can more effectively perform their functions in bipedal activities, and the elongation and forward convergence of the iliac crest extends the attachment to it of the abdominal muscles and reorientates them so that they are better able to support the abdominal viscera in the erect position and to balance the trunk on the hind limbs in side-to-side movements. The broadening of the ilium in man mainly affects its posterior region which is extended backward to give attachment to some of the strong muscles of the back that also play an important part in maintaining the trunk in a vertical position. This backward bending of the ilium, it will be noted, leads to the formation of a deep, angulated notch, the sacrosciatic notch, between it and the lower posterior bony element of the os innominatum, the ischium; the notch is bounded below by a prominent process, the ischial spine. A little way behind the anterior superior iliac spine at the front end of the iliac crest there is a well-marked thickening called the tubercle of the iliac crest; from this a strong bony ridge extends down the ilium to the articular socket (acetabulum) for the head of the thighbone. This ridge forms a sort of buttress that serves to strengthen the ilium in this position, taking the strain of some of the gluteal muscles. Neither the tubercle of the iliac crest or the buttress are to be found in the os innominatum of apes. Nor do the latter show (except to a very minor degree) the strong sigmoid curve of the iliac crest, as viewed from above, which is so characteristic of the human ilium, though quite variable in its development even in *Homo sapiens*. This curve is partly caused by the inward curving of the front end of the crest in man, and partly by an outward curving of the posterior end related to the greater width of the sacrum.

The acetabular socket faces laterally in the anthropoid apes, but with the rotation of the ilium in *Homo sapiens* it tends to face more forward. Just above the acetabulum in man is a stoutly built blunt eminence, the anterior inferior iliac spine. This has a functional importance of some significance, for to it is attached one of the extensor muscles of the knee, the rectus femoris muscle, and also a strong ligament, the ilio-femoral ligament, that braces the front of the hip joint when it is fully extended in the erect standing position. This bony process is absent or but faintly indicated in the large anthropoid apes.

Now let us turn attention to the ischial portion of the hipbone. At its lower end, and in apes situated at some distance below the acetabulum, is a roughened bony boss, the ischial tuberosity, on which the weight of the trunk is supported in a sitting position. It gives attachment to the hamstring muscles whose function it is to extend the thigh and flex the knee joint; in apes its surface faces mainly downward. In man, the tuberosity faces more backward in the erect position of the body, and it is situated quite close to the acetabulum. This difference in the relative position of the tuberosity has a functional significance. Its low position in apes, relative to the hip joint, enhances the leverage power of the hamstring muscles in extending the joint during movements of forward propulsion (as in quadrupedal mammals generally). In *Homo sapiens* such a leverage is no longer necessary since, in the habitually erect posture associated with full extension of the thigh, and in walking movements that involve no high degree of flexion, powerful, active extension of the hip joint by these muscles in forward propulsive movements is not required.

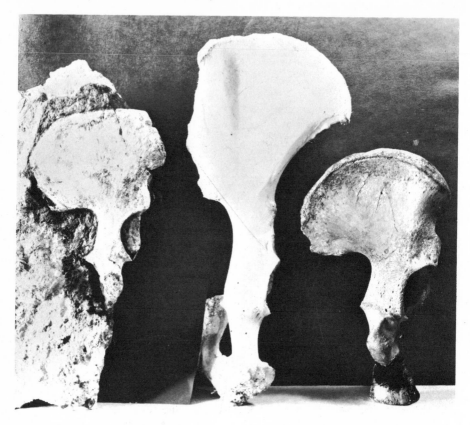

Figure 21. The photograph shows (left to right) the os innominatum of *Australopithecus africanus* found at Sterkfontein (still incompletely freed from the stalagmitic matrix in which it was embedded), the os innominatum of a chimpanzee, and that of a Bushman. (Courtesy of Broom and Robinson.)

On the inner aspect of the ilium near its posterior border is an articular surface for the sacrum, and it is through this sacroiliac joint of either side that the weight of the body is transmitted to the hind limbs. In apes the shape of the articular suface is an elongated, vertical, and narrow oval, and it is placed well above the acetabulum and at some distance from the latter. Its anterior margin is directed forward. In man it is relatively large and is placed lower down in relation to the acetabulum; it has also become rotated downward and forward so that the lower part of its anterior margin is disposed more or less horizontally. Finally, it should be mentioned that the junction of the pubic elements of the os innominatum, the pubic symphysis, is relatively longer in apes than in *Homo sapiens,* and in this respect is more similar to the pubic symphysis of quadrupedal mammals generally.

Attention must now be given to the sacrum that comprises the posterior wall of the pelvis as a whole, interposed between the iliac bones and articulating on either side with them. In apes it is long and narrow, in man short and broad. The greater breadth of the sacrum in man allows for a more extensive attachment on its posterior aspect of the powerful back muscles that are so important for maintaining the spinal column in a vertical position. The vertical position of the spinal column adapted for an erect stance leads to the formation in it of pronounced curvatures, including a forward cervical curvature of the neck vertebrae above, and a forward lumbar curvature below (see Fig. 20). In some apes a slight degree of lumbar curvature may be present (particularly in the gorilla), but it is never accentuated to the degree characteristic of *Homo sapiens.* Robinson has reported that the lumbar vertebrae found with the Sterkfontein pelvis give evidence of "a distinct lumbar curvature, as is readily demonstrated by articulating the vertebrae with the zygapophyses in contact."

Several remains of the australopithecine pelvis have so far been discovered in South Africa, one at Sterkfontein (practically complete except for the lower half of the sacrum), one at Swartkrans, and two immature specimens at Makapansgat. All of these show a typically hominid configuration, even though they also show a number of significant differences from the pelvis of modern man. Thus, the ilium is short in its vertical axis and broad from front to back. The iliac blade is extended backward, and the sacroiliac joint is orientated as it is in hominids. The sacrosciatic notch is deep and the roughened surface of the ischial tuberosity faces somewhat posteriorly. The sigmoid curve of the iliac crest is quite evident in some of the australopithecine specimens, for example, in those from Makapansgat and Swartkrans. The acetabular socket (at least in the gracile type of *Australopithecus*) is titled forward from the lateral direction characteristic of the apes. The pubic symphysis is relatively short and the sacrum is wide.

The features in which the australopithecine pelvis differs from that of *Homo sapiens* are quite marked and lead to the obvious inference that the australopithecines had not perfected erect bipedalism to the degree characteristic of modern man. In particular it is probable that they were not capable of the striding gait in walking but were evidently more capable of running in the erect position. The iliac blades are more widely

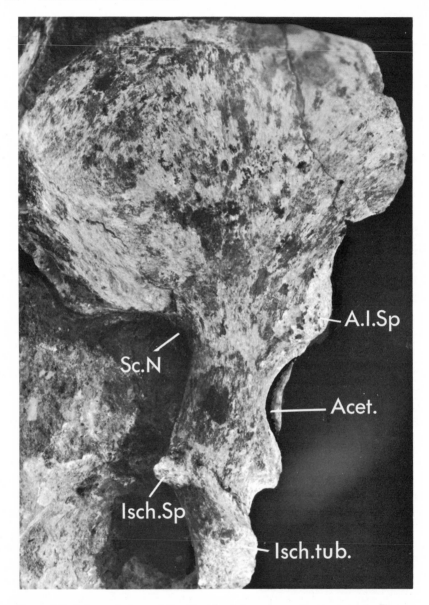

Figure 22. Enlarged view of the Sterkfontein os innominatum seen in Figure 21 to show the sacrosciatic notch, the ischial spine, the strongly developed anterior-inferior iliac spine, the acetabular cavity, and the ischial tuberosity. Note that the anterior-superior iliac spine is missing in this specimen.

splayed apart than in *Homo,* and in this respect (in *Australopithecus robustus* according to Napier) they were approximately intermediate between the gorilla and modern man [56]. The ilium is prolonged forward as a pointed process that is more obtrusive in *Australopithecus robustus* than in *Australopithecus africanus* (see Fig. 25). This process has disappeared in modern man, and Napier suggests that its disappearance in later hominid evolution is related to the stronger development of a lumbar curvature and the further rotation forward and inward of the iliac blade. He goes on to say, "The beak-like anterior superior spine . . . would appear to represent a morphologic compromise to ensure an upright posture and a bipedal gait in the absence of a fully medially rotated iliac crest. Mechanically, its effect is to project the anterior bony attachment of the abdominal muscles as far in front of the axis of the vertebral column as possible." The tubercle of the iliac crest, and the thickened buttress of bone that, in *Homo sapiens,* extends down to the acetabulum, are absent or only incipiently developed.

The tuberosity of the ischium is placed at a lower level in relation to the acetabulum, though not so much as in the large apes. But there seems to have been a considerable amount of variation in this feature. For example, in the somewhat mutilated os innominatum found at Swartkrans *(Australopithecus robustus)* the tuberosity is further removed from the acetabulum than in the pelvis of the gracile australopithecine from Sterkfontein, and in a Makapansgat specimen it is almost (but not quite) as near the acetabulum as in modern man. It may be inferred, therefore, that in the gracile type the functional use of the hamstring muscles in the erect gait must have approximated much more closely to that of *Homo.* The area of the sacroiliac joint is considerably smaller than in modern man, suggesting that the transmission of the weight of the body through this joint to the lower limbs was less effectively developed.

There is one significant feature of the australopithecine pelvis to which attention does not appear to have been given by other writers. In the modern human ilium there is a broad groove situated just medial to the anterior inferior spine (see Fig. 23). This groove is formed by a powerful flexor muscle of the hip joint, the ilio-psoas muscle, which extends down from the lumbar vertebrae and the iliac fossa, and, crossing over the brim of the pelvis, is then deflected downward and backward to reach the upper end of the shaft of the thighbone. In apes, in which the thigh is not usually held fully extended in an habitually erect posture, the backward

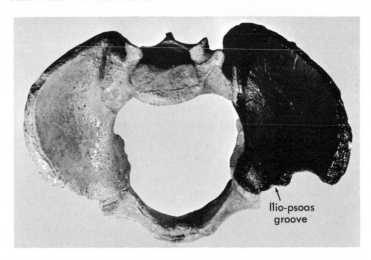

Ilio-psoas
groove

Figure 23. The left iliac bone of an immature australopithecine
found by Professor Raymond Dart at Makapansgat and here shown
from above articulated with an adolescent Bushman pelvis. The
ilio-psoas groove is well shown. (Courtesy of Dart.)

deflection of the ilio-psoas muscle as it passes to its insertion into the
thighbone is not so pronounced; consequently no well-marked groove is
made by the muscle at the brim of the pelvis. This groove is quite well-
marked in the australopithecine pelvis, and this argues strongly for a full
extension of the thigh in an habitually erect stance.

We may now summarize the significant features of the australopithecine pelvis: hominid construction marked by the relative breadth of the
ilium; the backward extension of the posterior end of the iliac crest and
the low position of the sacroiliac articulation in relation to the acetabulum; the orientation of the sacroiliac articulation relatively to the longitudinal axis of the os innominatum; the sharply angulated sacrosciatic
notch associated with a prominent ischial spine; the strongly developed
anterior inferior iliac spine; the orientation and position of the ischial
tuberosity in relation to the acetabulum (particularly in the Makapansgat
pelvis); the well-marked groove at the brim of the pelvis for the ilio-psoas
muscle; and the relative breadth of the sacrum. In all these characters,
even when taken individually, the pelvic skeleton makes a strong contrast
with the modern anthropoid apes (see Fig. 24); taken in combination,
they comprise a total morphological pattern that is distinctive of the

Figure 24. Front view of the pelvis of (A) a chimpanzee, (B) the Sterkfontein australopithecine, and (C) a Bushman. Note that in the australopithecine pelvis the iliac blades are more widely splayed apart than in the Bushman, but not so widely as in the anthropoid ape; the sacrum is relatively wide and the pelvic symphysis relatively short. (B and C: Courtesy of Robinson.)

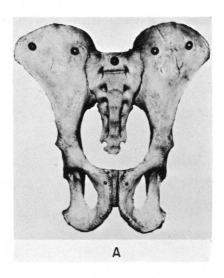

A

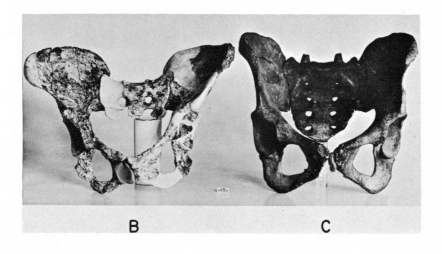

B   C

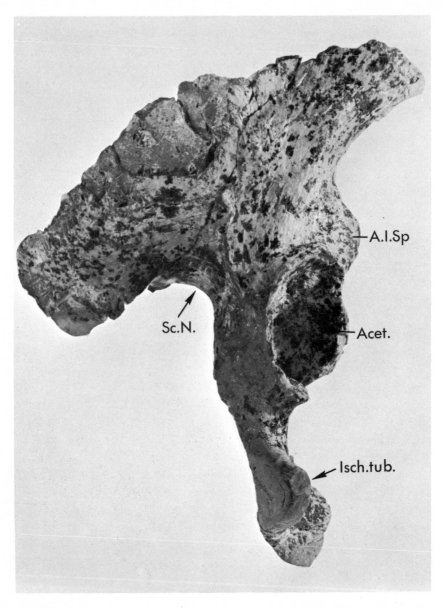

Figure 25. The os innominatum of *Australopithecus robustus* found at Swartkrans. Note the deep sacrosciatic notch (Sc.N.), the strongly developed anterior-inferior iliac spine (A.I.Sp.), and the low position of the ischial tuberosity (Isch. Tub.). This specimen is incomplete (most of the iliac crest is missing) and the acetabulum and ischial element have been distorted by crushing.

Hominidae among all other groups of mammals. Moreover, they are characters that are quite certainly related to posture. The broad ilium extends the antero-posterior attachment of the buttock muscles that are used for balancing the trunk on the lower limbs; the bending down of the posterior extremity of the iliac crest brings one of the muscles (the gluteus maximus muscle) to a position behind (instead of lateral to) the hip joint and so permits it to play its essential role as an extensor of the hip joint in walking erect; the approximation of the sacroiliac articular surface to the acetabulum makes for greater stability in the transmission of the weight of the trunk to the hip joint; the reorientation of the sacroiliac articulation is consequent on a rotation of the sacrum, which is associated with modifications in the disposition of the pelvic viscera; the robust anterior inferior iliac spine serves in part to attach the powerful ilio-femoral ligament which braces the front of the hip joint in full extension in the erect standing position; the deep groove for the ilio-psoas muscle is related to the fact that in the habitually extended position of the thigh in the erect posture the muscle has to turn backward at a marked angle to reach its insertion onto the thighbone; the broad sacrum provides for the attachment of powerful back muscles that are required for maintaining the spinal column in a vertical posture. From considerations such as these it is a reasonable, and indeed an inevitable, inference that the australopithecines had become adapted to an erect bipedalism. This is not to belittle the equally important inference to be drawn from pelvic structure that the australopithecines had not perfected the mechanism for erect bipedalism that later was finally achieved in the genus *Homo*. As we shall see, these general conclusions that the australopithecines were habitual bipeds of hominid status, but, particularly in the achievement of a striding gait and the mechanics of weight transmission to the lower limb were not such efficient bipeds as modern *Homo sapiens,* are also the conclusions that have been drawn from a study of the femur, the bones of the lower leg (tibia and fibula), and also the foot bones found at Sterkfontein and Olduvai. In other words, the accumulated evidence derived from all these remains is quite consistent within itself.

FUNCTIONAL IMPLICATIONS
OF AUSTRALOPITHECINE
LIMB BONES

In a previous chapter I have made brief references to some of the limb bones of *Australopithecus,* partly in order to give a general preliminary picture of this extinct genus of hominids, and partly to make it clear that in my opinion those found in the deposits of the Olduvai Gorge in Tanzania belonged to australopithecines and not to the genus *Homo* (as has been claimed). In the present chapter I propose to examine the anatomical details and functional implications of the limb bones in greater detail. In the monograph published by Dr. Broom early in 1946, he described and figured a few bones of the upper and lower limbs that he had found in association with australopithecine skulls at Sterkfontein and Kromdraai, and these I was able to study during my first visit to South Africa at the end of 1946.

The lower extremity of the upper armbone or humerus and the upper end of one of the forearm bones, the ulna, found at Kromdraai, show a close resemblance to those of *Homo sapiens,* and if they had been found alone and isolated from any other fossil remains would almost certainly have been ascribed to the genus *Homo*. They probably both belonged to one individual, for they are both of the right arm and articulate perfectly with each other. They are relatively small and slender in build. The humerus lacks the strong ridge on its outer side, the supinator ridge, which is commonly well developed in the modern large apes for the strong muscles that are used by these creatures in their climbing activities. On the back of the bone the olecranon fossa that accommodates the upwardly projecting olecranon process of the ulna in full extension of the forearm corresponds in its extent and depth to that of man, and thus contrasts with the rather deep excavations seen in chimpanzees and gorillas (associated with the powerful development in these apes of the olecranon process). On the other hand, the position of the articular surface for the other forearm bone, the radius, is not set quite so far forward as it normally is in *Homo sapiens,* and the alignment of the trochlear articular surface for the ulna is also different. Mechanically this has been taken to indicate some limitation of the power of flexion and a capacity for hyperextension at the elbow joint to a degree not usually possible in recent anthropoid apes or man. Straus [85] has pointed out it is not always easy to distinguish between the lower end of a human humerus and that of a chimpanzee (and this is particularly the case with small and lightly built females), so that too much weight should not be given to the apparently hominid characters of the Kromdraai humerus to which I have called attention. The upper end of the ulna is certainly not distinguishable from that of man, and it contrasts with that of the large anthropoid apes in which the olecranon process is usually massively developed with particularly prominent ridges on its outer and inner aspects for the attachment of powerful arm muscles.

One of the anklebones, the talus, which was found at Kromdraai, is small and presents a number of features indicating a general construction in some ways intermediate between that of *Homo sapiens* and pongids (see Figs. 26 and 27). The upper joint surface (which articulates with the tibia) resembles that of man in being broad in relation to other parts of the bone and also in its even curvature from side to side. In the gorilla and chimpanzee the surface is commonly somewhat flattened toward its inner side, and to its outer side it slopes gradually upward toward its outer border. This asymmetrical contour in apes is possibly associated with the fact that the weight of the body is transmitted through the lower limbs more to the inner side of the talus. In man and *Australopithecus,*

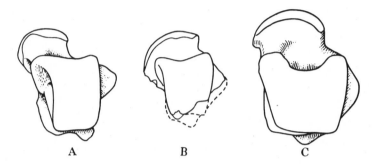

Figure 26. The upper aspect of the right talus of (A) a chimpanzee, (B) *Australopithecus robustus* (found at Kromdraai), and (C) *Homo sapiens*. (*Journal of Anatomy*, Vol. 81, p. 323.)

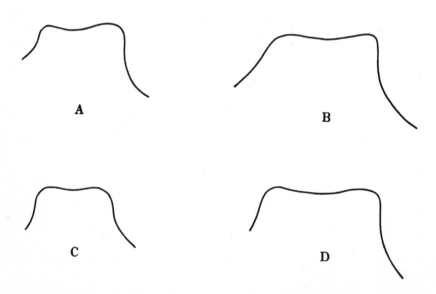

Figure 27. The contours of transverse sections through the talus in the plane of the maximum extension of the medial and lateral facets of (A) a chimpanzee, (B) a gorilla, (C) *Australopithecus*, and (D) *Homo sapiens*. In all cases the lateral surface of the bone is to the right. (*Journal of Anatomy*, Vol. 81, p. 323.)

on the other hand, the shape of the upper articular surface indicates a more even distribution in the transmission of the body weight to the talus. In the human and australopithecine talus, also, the vertical extent of the articular facet on the outer surface for the fibula is shorter in relation to the width of the upper tibial articular surface. The articular surface for the tibia on the inner aspect of the Kromdraai bone is quite similar in contour and extent to that of man and is approximately vertical in its disposition; in the chimpanzee and gorilla it slopes inward to a marked degree (Fig. 27). The neck of the fossil talus is short and broad, and this is associated with an unusual inward extension of the joint surface of the head of the bone which articulates with one of the tarsal bones of the foot in front (see Fig. 28). Although the contour and slope of the articular surfaces on the body of the bone indicate that the latter was constructed for stability in weight-bearing, and for a more efficient hinge-joint mechanism in walking and running than is to be found in the anthropoid apes, it has been suggested that the inward extension of the articular surface of the head, perhaps, allowed for a greater deflection inward of the body weight to the forefoot in association with a mobile big toe capable of some degree of divergence from the other toes. This may have been so in the case of *Australopithecus robustus* whose remains were found at Kromdraai, but that it was evidently not so in all australopithecines has been made evident from the almost complete foot skeleton found at Olduvai. At any rate it is clear from the Kromdraai talus that the direction and weight transmission to the forward part of the foot in the australopithecines differed from the condition in modern man.

The discovery of the Olduvai foot skeleton was reported by Dr. Leakey in 1960, and its details were briefly described by Dr. Day and Dr. Napier four years later [27]. The skeleton is complete except for the phalanges of the toes, the heads of the metatarsal bones, and the posterior part of the heel bone or calcaneus (Fig. 28). The foot as a whole is small and is stated to show a remarkable resemblance in its anatomical details to that characteristic of *Homo*. Among other details, there is a joint facet between the bases of the metatarsal bone of the big toe and that of the second toe making it clear that there was no divergence between these toes as in the anthropoid apes. It does show significant differences from the foot skeleton of modern man. For example, the metatarsal bone of the middle toe is unusual in its length and thickness, and Day and Napier suggest that this may be taken to indicate "that the *Homo sapiens* pattern of metatarsal structure which serves the transmission of weight and pro-

Figure 28. Upper aspect of the foot skeleton of one of the australopithecines found at Olduvai in Tanzania. The front ends of the metatarsal bones (above) and the back part of the calcaneus (below) are missing. (Reprinted from *The Fossil Evidence for Human Evolution,* 2d ed., by Sir Wilfrid E. Le Gros Clark, © 1955 by the University of Chicago, by permission of the University of Chicago Press.)

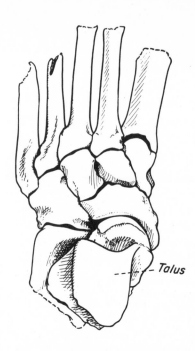

Talus

pulsive effort through the forefoot had not fully evolved." They also point out there are certain features of the talus that are not characteristic of *Homo sapiens,* and they particularly note certain points of similarity with the australopithecine bone from Kromdraai, for example: "The horizontal angle and the angle of the neck are similar in both forms." They go on to list the skeletal adaptations in the foot of modern man associated with a striding bipedal gait and conclude that in the Olduvai specimen "the tilt of the talus and the metatarsal robusticity pattern alone suggest that the unique striding gait of *Homo sapiens* had not yet been achieved." It will be observed, then, that the functional implications of the Olduvai foot skeleton parallel those of the australopithecine pelvis, and there is in fact nothing to suggest that it was not compatible functionally with the pelvis found at Sterkfontein. Indeed, they are in marked conformity with each other in this respect and there thus seems no reason to ascribe the Olduvai foot to the genus *Homo* (and the species *Homo habilis*) as some have done. On the contrary, the most reasonable inference is that it belongs to the genus *Australopithecus,* and perhaps even to the species *Australopithecus africanus* whose remains had been previously discovered in South Africa.

The two bones of the lower leg, the tibia and fibula, were found at

Olduvai on the same living floor as the almost complete skull of a representative of *Australopithecus robustus*. They have been described by Dr. P. R. Davis [26] and, like all the other australopithecine limb bones so far discovered, they show an interesting combination of traits characteristic of *Homo sapiens* and primitive hominoid traits. The lower ends of these fossil bones are very similar to those of modern man; for example, the joint surface of the tibia that articulates with the talus is orientated almost at a right angle to the longitudinal axis of the shaft in contrast to the lateral slant of the surface in the large anthropoid apes. Thus, as Davis remarks, "the region of the ankle joint shows itself to have been adapted for plantigrade locomotion," a condition of erect bipedalism. The nature of the curvature of the lower end of the shaft is also similar to that of modern man and differs from that of the pongids. On the other hand, in the relative extent of the areas for two of the muscles attached to the back of the tibia there is a closer resemblance to the apes. At the upper end of the back of the tibial shaft the marking for the short muscle behind the knee joint, the popliteus muscle, is stated to differ widely from both the modern human condition and that of the modern apes, and the extent of the attachment of one of the calf muscles, the soleus, appears to have been intermediate between these two types. Davis concludes that the tibia and fibula indicate that bony adaptation to bipedalism was well advanced at the ankle. He also states: ". . . it appears possible that the knee joint was less well adapted to bipedalism than it is in *Homo sapiens*. Thus while the fossil form was clearly a habitually bipedal plantigrade primate, its gait may well have differed considerably from that of modern man." Here, again, the functional implications of the Olduvai leg bones are in accord with those inferred from a study of the australopithecine pelvis, the talus from Kromdraai, and the Olduvai foot skeleton.

Now let us turn our attention to the thighbone or femur. Dr. Broom found two lower extremities of this bone at Skerkfontein in the same deposits that contained many remains of the gracile type of *Australopithecus,* and both are very similar in their anatomical configuration. The first specimen was found in 1937 and described by Broom in 1946 [14]. I had the opportunity of studying the original fossil in 1946 and took note of its remarkably human characters. The bone is small (about the size of the thighbone of modern Bushman). One of its striking features is the obliquity of the shaft, for the angle between the shaft and the vertical axis of the trunk in the standing position makes a strong contrast with that of the femur in the modern large apes (Fig. 30). The angle of obliquity is at

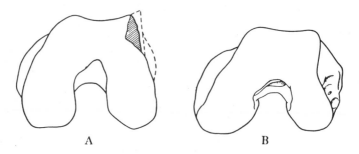

Figure 29. The lower end of a left femur of (A) *Australopithecus africanus* seen from below, compared with that of (B) an adult chimpanzee. Note the depth and relative narrowness of the intercondyloid notch in the australopithecine bone, and also the curvature of the articular condyles. (*Journal of Anatomy*, Vol. 81, p. 328.)

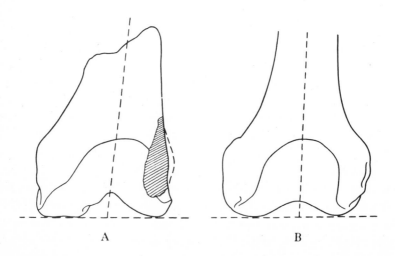

Figure 30. The lower end of (A) an australopithecine left femur found at Sterkfontein seen from the front, compared with that of a chimpanzee. Note the marked obliquity of the shaft in *Australopithecus*, and also the relatively greater robusticity of the shaft as compared with the ape femur. (*Journal of Anatomy*, Vol. 81, p. 327.)

least 7°, compared with a range of 4-17° in normal modern male English thighbones. On the other hand, the obliquity of the shaft in the African anthropoid apes is in most individuals much slighter. It has been inferred from this character that in the erect standing position adopted by *Australopithecus africanus* the femur sloped downward and inward as in man, and the contour of the articular surface for the patella (kneecap) is in conformity with this conclusion. The shape, curvature, and relative sizes of the condyles of the femur (which articulate with the tibia at the knee joint) are again typically human. They are separated by a narrow intercondyloid notch, in comparison with the wider notch in apes (Fig. 29). In the Sterkfontein femur the notch is prolonged forward and shows an impression made by contact with a ligament inside the knee joint, the anterior cruciate ligament, when the joint is held in full extension. Such an impression is typical of the human femur and appears to indicate that the australopithecine knee joint could be sustained habitually in an extended position in an erect posture as in modern man. But the intercondyloid notch is prolonged forward to an unusual extent, further indeed than in *Homo sapiens*. The significance of this feature is difficult to assess, but it does indicate a difference in the mechanism of the knee joint and is thus in conformity with the inferences drawn from the upper end of the Olduvai tibia. The second specimen of the lower end of the femur from Sterkfontein has been stated by Broom and Robinson [8] to agree in all essential respects with the first specimen, including the obliquity of the shaft of the bone and the unusual extension forward of the intercondyloid notch. The fact that there is a marked obliquity of the shaft in two successively found specimens of *Australopithecus* adds emphasis, of course, to its functional implications in relation to erect bipedalism, the more so when this character is considered in combination with the other hominid features of the lower end of the femur.

Unfortunately, very little is known of the upper extremity of the australopithecine femur. Dr. Broom found a badly preserved specimen in association with the pelvis at Sterkfontein but, beyond noting certain minor differences from *Homo sapiens* that suggested to him a somewhat different orientation associated with a similar difference in the orientation of the acetabular socket of the os innominatum, he gave no detailed description of the fragment. So far as can be ascertained from the damaged head of the bone, the size of the latter approximated to that of a Bushman. Two other upper femoral fragments from Swartkrans, but only doubtfully referable to *Australopithecus,* have been briefly described on the basis of plaster casts of the original specimens by Napier [56]. He remarks on the small absolute and relative size of the head of the bones, its transverse diameter falling outside that of modern man, and on the length of the femoral neck which exceeds that of any human or anthropoid femur that he had studied. This last character is perhaps related to the lateral orientation of the acetabular socket of the pelvis, and permitted wide side-to-side movements of the thigh so as to allow the legs and feet to be brought into close alignment in standing and walking. Other smaller differences from modern man to be noted relate to two bony eminences for the attachment of muscles descending from the pelvis, the greater trochanter which is rather small, and the lesser trochanter which is directed more backward. Finally the roughened line on the front of the femur at the junction of the neck and the shaft, the ilio-femoral line to which is attached the ilio-femoral ligament, is absent. But in the femur of modern man this may sometimes be but faintly marked.

Of considerable importance because of their functional implications for manipulative capacities are the hand bones of the australopithecines. These bones comprise some of the carpal or wristbones, metacarpals, and phalanges. Those readers who may not be well acquainted with their nomenclature may refer to Figure 31 which shows in outline the skeleton of a modern human hand as viewed from the back. One of the wristbones, the capitate bone, was found by Broom embedded in the stalagmitic matrix at Sterkfontein that had yielded many specimens of *Australopithecus africanus* (Fig. 32). As Broom pointed out, it is unusually small and in general has a very human appearance; this is shown particularly in the proportions of the bone, for it is shorter and relatively broader than that of the chimpanzee and gorilla (in both of which it is elongated in conformity with the elongation of the hand as a whole), and also in the shape of the lower articular surface. In the chimpanzee the latter is divided into several concave facets separated by intervening ridges, which articulate with the rather complex joint surfaces at the base of the third metacarpal bone. In the gorilla it is distinctly convex from side to side in its dorsal part for articulation with a concavity on the base of the metacarpal bone. In modern man it is much more flattened. In the Sterkfontein capitate bone this articular surface is gently curved as in man. In the fossil bone, again as in man, there is an elongated facet for the second metacarpal which is directed largely downward; in the gorilla and chimpanzee this facet is relatively smaller and broadly rectangular in contour, and it faces directly outward on the outer surface of the bone, being separated from

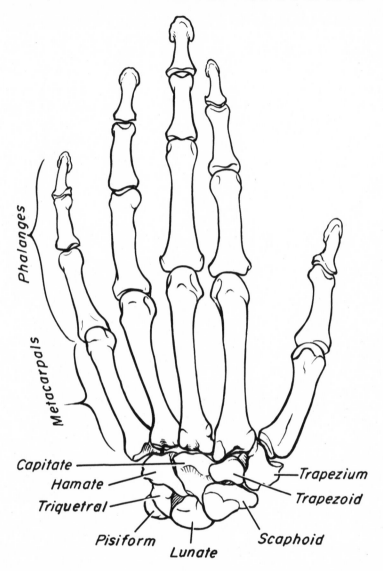

Figure 31. Diagram of the skeleton of a modern human hand indicating the names and positions of its various bony elements.

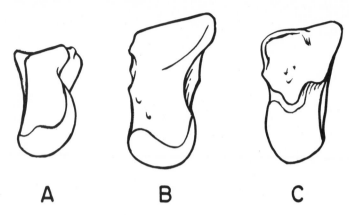

A   **B**   **C**

Figure 32. The capitate bone of (A) *Australopithecus africanus* found at Sterk-fontein, compared with that of (B) a chimpanzee, and (C) modern man. All are viewed from the posterior aspect.

the facet for the third metacarpal by a sharp acute-angled border. A feature of some importance in the fossil bone is the presence of a pronounced concavity in the middle of its outer aspect, and this gives it a somewhat "waisted" appearance when viewed from in front (Fig. 32). In the chimpanzee and gorilla this concavity is a still more pronounced feature and presumably serves to attach a strong interosseous ligament binding it to the adjacent trapezoid bone of the carpus. In the capitate bone of *Homo sapiens* the outer surface is usually flattened or broadly convex so that the whole bone appears more robust. Lastly, a primitive hominoid feature is presented by a rough, bluntly pointed, process that projects forward from the front surface of the bone. This process is but faintly indicated in modern man, but it is even more strongly developed in the chimpanzee. Thus, although the Sterkfontein bone is predominantly hominid in its main features, it does appear to occupy a position somewhat intermediate between *Homo sapiens* and the large anthropoid apes in some of its anatomical characters. In size the fossil bone is equivalent to that of a Bushman and very much smaller than that of an adult chimpanzee. Clearly it belonged to a small hand and is suggestive of a small and lightly built individual; in this respect it conforms quite well with the small and delicately built pelvis that was also found at Sterkfontein. The extent and disposition of its articular surfaces, and of the roughened surfaces for the attachment of interosseous ligaments, may be taken to indicate that the movements between the capitate and the adjacent bones of the hand were less free than in modern man, but much more free than in the hand of the modern anthropoid apes.

Two metacarpal bones found at Swartkrans have been described by Napier [53]; one is a complete metacarpal of the thumb, and the other the lower half of (probably) a fourth metacarpal. According to Napier the thumb metacarpal is short, corresponding to the shortest metacarpal in a series of Bushman bones; it is also an unusually robust bone but the extent and degree of its muscular markings "suggest a closer affinity with the human pattern of muscular activity than with that of the anthropoid apes." Napier also points out a resemblance to the gorilla in the extent of the joint surface at the base of the bone, which articulates with the trapezium bone of the wrist, and he also notes a similarity to the modern large anthropoid apes in the marked longitudinal curvature of the shaft, and in the shape, curvature and obliquity of the head of the bone which articulates with the proximal phalanx of the thumb. Nevertheless, he concludes from the apparent mobility of the articular surfaces that the hand was not only capable of prehensile functions as in man, but possessed an extremely powerful grip. The fourth metacarpal is stated to be of slighter build and to approximate more closely to *Homo sapiens* than the thumb metacarpal, and Napier suggests by implication that it belonged to a more advanced hominid that occupied the Swartkrans site, for example to the type originally referred to as the genus *Telanthropus*, and later to a pithecanthropine, *Homo erectus*. But we have already expressed serious doubts about this interpretation based on such few and fragmentary fossil specimens, taking the view that they merely represent the remains of a small and lightly built individual of the genus *Australopithecus*. Incidentally, it is not a little remarkable that whenever skeletal remains have been found in australopithecine deposits that are smaller and more lightly constructed than the majority of the australopithecine fossils, it has been so commonly assumed that they are the remains of a more "advanced" hominid genus without taking into consideration that australopithecine individuals (particularly of different sexes) presumably varied as much in size, form, and robust build as those of other hominoid species. In fact, Napier reported that in the fourth metacarpal bone from Swartkrans, "The strong interosseous markings of the fossil bone suggest the presence of powerful muscles and therefore of divergent metacarpals and of a broad span to the hand," and he also stated that the dorsal interosseous crest (to which is attached some of the small intrinsic muscles of the hand) is more sharply defined than in any of the modern human bones that he examined.

Now let us consider the hand bones found at Olduvai. Fifteen of these bones were discovered, belonging to at least two individuals. These include only three carpal bones, all damaged. According to Napier [54] the capitate bone is in general more hominid than apelike (agreeing in this respect, therefore, with the Sterkfontein capitate bone). The articular surface of the trapezium for the first metacarpal is saddle-shaped, indicating an opposable thumb that was possibly capable of a "precision grip" allowing of finer manipulative activities, but "the evidence provided by the other articular surfaces indicates that its 'set' in the carpus was unlike that found in modern man and similar to the condition in *Gorilla*." This gorilloid feature, it is important to note, is concordant with the gorilloid feature previously observed by Napier in the articular surface at the base of the thumb metacarpal from Swartkrans. The other hand bones from Olduvai are phalanges. They are all robustly constructed, with a marked dorsal curvature, and show strong impressions and ridges that were evidently associated with powerful flexor muscles of the fingers. The terminal phalanx of the thumb is consistent with the inference as to the opposability of this digit—a degree of opposability that is a distinctive feature of the Hominidae in contrast to the modern Pongidae.

I have drawn attention to the conformity of the Olduvai carpal bones with the capitate bone of *Australopithecus* from Sterkfontein and with the inferences drawn from the thumb metacarpal of *Australopithecus* found at Swartkrans. The robust and strongly curved phalanges also conform remarkably well with the robust and strongly curved Swartkrans metacarpal. It may well be questioned, then, what plausible reason could there have been for assigning the Olduvai bones to another genus and species altogether—*Homo habilis*—particularly as they were found on the same living floor as two fragments of the skull indicating a very small cranial capacity within estimations for the probable population range of *Australopithecus* and well below that of any known type of *Homo*. It seems to me that there is no sound reason for making such a nomenclatural contrast. And the evidence for the australopithecine affinities of the Olduvai hand bones and cranium applies equally well, as we have seen, to the Olduvai foot and leg bones. Indeed, all these sources of evidence are mutually concordant.[1]

---

[1] A clavicle (collarbone) has also been found at Olduvai. No detailed report of this bone has to date been published, but in the report of a discussion in *Current Anthropology* (Vol. 6, 1965, p. 402), Napier has stated that it is like that of *Homo sapiens* but distinguishable by certain anatomical details.

In summary, we may reach certain rather positive conclusions regarding the australopithecines so far as their limb structure is concerned. Every authority, whether anatomist or palaeontologist, who has expressed an opinion on the matter in recent years, is agreed that they had already acquired an habitual erect bipedalism in gait and stance based on a modification of the hind limb that is characteristic of the Hominidae (and very different in this respect from the Pongidae). It is also agreed that the mechanism for bipedalism at that phase of hominid evolution had not been perfected to the degree achieved by modern man. In particular, as Napier and others have stressed, the australopithecines were not capable of the unique "striding gait" that characterizes *Homo sapiens,* and that probably characterized the extinct types of the genus *Homo* so far known. They may have walked with a sort of shuffling gait, but they were capable of an efficient running gait. The evidence for these inferences, as I have pointed out, is based on the study of the foot skeleton and other bones of the hind limb. As we shall see, there is good reason to believe they hunted animals for food, and their running facilities would have enabled them to pursue their prey very effectively. The skeletal elements of the australopithecine arm and hand also make it clear that their upper limbs were not specialized for arboreal activities as they are in the modern large anthropoid apes; on the contrary these give evidence of a mobile arm and of a hand with an opposable thumb that was capable of the finer manipulative movements required for holding and grasping small objects. Their hands would certainly have allowed them to use tools of some sort, and there is no reason to suppose that they were not capable of fabricating crude tools, whether of stone or bone. The inference that they were indeed capable of making tools is further provided by the indirect evidence of undoubted implements that have been found closely associated on living floors with their skeletal remains.

HOW DID
THE AUSTRALOPITHECINES
LIVE?

It is clear from the study of their fossil remains that, anatomically speaking, the australopithecines were peculiarly defenseless creatures. They lacked the sharp and formidable canine teeth characteristic of the anthropoid apes that could be used for attack or for protection against attack. And, although they certainly varied in size, many of them, the gracile type of *Australopithecus* in particular, were small and light in build. Yet in South Africa they inhabited regions of relatively arid veldt country apparently closely similar in climate to that of today. There were indeed some fluctuations in climate during the occupation by the australopithecines of the Transvaal, as indicated by the different animal remains found in association with them [29], some of which are today savannah-living types or even adapted for desert conditions. Others, such as the hippopotamus whose remains have been found in the australopithecine deposits at Makapansgat are taken to indicate adjacent rivers of larger size than exist today. Periodical changes in rainfall and relative humidity have also been inferred from geological studies of the deposits, for example, Dr. Brain's arduous analysis [3] of the proportional amounts of quartz and chert grains which they contain ("an increase in the proportion of foreign quartz grains . . . tends to indicate drier conditions associated with more intense wind action"), and his estimations of the varying percentage of wind-abraded sand grains at different stratigraphical levels. From such evidence as this it has been concluded that the gracile australopithecines at Taung, Sterkfontein, and Makapan lived during a period of minimum rainfall, while the robust type at Kromdraai and Swartkrans lived during a period of increasing rainfall. But, in spite of these successive variations in relative aridity and relative humidity, it is certain that the australopithecines did not live in regions of the sort of tropical forest for which the modern anthropoid apes are adapted. They occupied an environment where they could hardly be expected to survive the dangers of predatory animals unless they were able to move rapidly in open grasslands or wooded savannah country and were capable of using or fabricating weapons of some kind. As Bartholomew and Birdsell have remarked [2], in considering the likely ecology of the australopithecines, and taking into account their small canines and incisors and nonsectorial premolars, "These dental characteristics are unique to them among all large carnivorous mamals," and they take this to imply the killing of game by simple tools.

Is there any reason to assume that the australopithecines not only used tools but made them? On this question I am not, as an anatomist, able to write with full authority, but I have tried to assess all the evidence objectively, and I feel assured that the answer to both these questions is Yes. Here we must be clear about the distinction between tool-using and tool-making. Tool-using is not uncommon among mammals, and even birds. As examples, I may mention the habit of sea otters of using a pebble to crack open the shell of sea urchins, or the habit of some of the finches in the Galapagos Islands of using thorns of appropriate size held lengthwise in their short beaks for prying out insects and grubs from crevices in the bark of trees. There is also the interesting observation reported by Dr. Vevers and Dr. Weiner [91] on a small captive capuchin monkey that deliberately, and without any training, selected a marrow bone as a ready-to-hand tool for cracking walnuts. It seems likely enough, therefore, that the australopithecines could, and did, make use of similar ready-to-hand tools for similar purposes. But chimpanzees in the wild are not only tool-users, for, as observed by Miss Goodall [30], these apes can actually fabricate tools in the sense that they will deliberately select twigs for probing into termite hills in order to get at the ants, breaking the twigs into lengths suitable for their purpose and even trimming off side branches if these obstruct the passage of their probing implements. Some authors would refer to this procedure as "tool-modifying" rather than "tool-making," but to my mind this seems to be a distinction without any clear-cut difference. Incidentally, it is interesting to note that chimpanzees have been seen to wander away from the immediate proximity of a termite hill in order to collect a number of appropriate stalks all of which they take back with them for use one by one. Surely this indicates some degree of premeditation; indeed, Miss Goodall goes so far as to say that "fishing" for termites with twigs is not an inborn behavior pattern; it is a social tradition which represents the emergence of a primitive culture, if culture consists of "behaviour patterns transmitted by imitation or tuition." Nevertheless, it is as well not to overemphasize these rudimentary tool-making incidents that fulfill the need for some immediately visualized requirement; there is a wide gap between the mental ability to undertake such simple tasks and the conceptual capacity of the human mind that makes possible the fabrication of implements for *future* contingencies.

Let us now return to the australopithecines. It is known that crude stone implements were being fabricated in Africa in the early part of the Pleistocene when they were living there, for implements of this sort have been found in river gravels that were deposited at about the same time. More than this, roughly prepared stone tools have actually been found in association with australopithecine remains at Swartkrans, Sterkfontein, and Olduvai. But a number of archaeologists have doubted whether they were, in fact, actually fabricated by the australopithecines themselves rather than by some hypothetical more advanced hominid of pithecanthropine status that coexisted with them. Such an interpretation is largely based on negative evidence—the absence of any stone implements in some of the cave deposits containing skeletal relics of *Australopithecus*. But it has also been questioned whether the australopithecines did live in these caves. The latter may have been the lairs of large carnivores that preyed on the australopithecines and left some of their remains there. This is a plausible hypothesis, for it has been argued that such defenseless creatures as the australopithecines would hardly have lived in the deeper recesses of caves where they could easily have been cornered by carnivores unless they knew the use of fire to frighten them off. And there is not the slightest evidence that they had learned the use of fire. It is more likely, therefore, that they customarily occupied rock shelters rather than caves at night, with guards to warn of the approach of predators. If such were the case, they might have made and used stone implements elsewhere, for example in open ground during the day, or in their rock shelters. This, then, may account for the implements that have been found at some of the australopithecine sites. Another argument has been based on the assumption that with such small brains they could not have had the mental capacity for fabricating implements, however crude in construction, by deliberately chipping stone pebbles to form sharp-edged tools. But the fact is that, though we know the australopithecine brain was quantitatively small, we know nothing of its qualitative functions except from the indirect evidence of some of the endocranial casts that the gray matter of the cerebral cortex was quite complexly convoluted. The mere mass of the brain is not so important for assessing intellectual abilities as the organization of its intrinsic structure. Even in *Homo sapiens* the extremes of cranial capacity in individuals range from 900 cc, or even less, to almost 2000 cc, without in every case any evident difference in intelligence. Napier concluded from his studies that the australopithecine hand was certainly capable of what he calls a "power grip," that is, the sort of grip that we use in wielding a hammer, and that they may also have been capable of a "precision grip," the sort of grip for holding small objects by opposing the thumb and fingers to each other. The same author [54] has also demonstrated by practical experiment that the power grip alone is adequate for constructing not only the crude stone tools called "pebble tools," but even more advanced types of "hand ax" similar to one of the tools found at Sterkfontein. Thus we can say that the australopithecine hand was capable of the manipulative movements that would have been required for making the Sterkfontein stone tools and on this account, at any rate, there is no need to postulate for their manufacture the presence of an unknown hominid of the pithecanthropine type.

As regards the implements found at Swartkrans, the suggestion has been advanced that these were made by the owners of the smaller jaws and teeth found at this site, which were at one time assigned to a new genus *Telanthropus* and later to a pithecanthropine *(Homo erectus)*. But, as I have argued in a previous chapter (Chapter 4), there seems no adequate reason to suppose that these very fragmentary and incomplete fossils were anything more than the relics of an unusually small and lightly built individual of the australopithecine community occupying the Swartkrans site.

No doubt the most striking evidence of the tool-making capacities of the australopithecines was provided by Dr. Leakey when he discovered a skull of the robust type of *Australopithecus* at Olduvai. He reported [41] that it was found on a living floor strewn with crude stone implements and the flakes that had evidently been struck off in trimming them to form sharp edges, and many of the tools had been constructed of material that is not indigenous to the site but must have been brought there from some distance; he concluded, naturally enough, that the skull represents one of the hominids who occupied the living side and who made and used the tools associated with it. His discovery in 1963 of the jaws and cranial fragments which he and his colleagues referred to a species which they christened *Homo habilis* led him to doubt his previous assumptions, for he now suggested that the tools were after all not in fact made by australopithecines, but by this supposedly more "advanced" hominid. I have already expressed with supporting arguments my opinion (which is also the opinion of a number of reputable authorities) that the so-called *Homo habilis* is a representative of the genus *Australopithecus*.

So far as I am aware, it has never been explicitly denied that the australopithecines could, or did, fabricate crude implements of stone. For my part, I find the negative evidence against such a conclusion far from convincing, while the positive evidence of the association in the same deposits of stone implements mixed up with australopithecine remains (and with no commanding evidence of the contemporaneous existence at the same sites of more advanced hominids) is far more conclusive.[1] In any case, as I have already remarked, if the australopithecines did not make and use weapons and implements of some kind, how could they possibly have survived all the hazards of the predatory animals among which they lived?

It is now generally agreed that the Australopithecines were hunters, and that their diet included animal food. The evidence for this conclusion rests mainly on the large quantities of animal bones of various kinds, some evidently split open for their marrow content, that have been found in closely packed collections at the australopithecine sites. It has been argued, to and fro, whether these collections of bones may not have been the result of the predatory activities of carnivores that from time to time occupied the caves containing australopithecine remains, or of the scav-

[1] One writer, who appeared to doubt whether some of the stone artifacts found with the remains of australopithecines were actually fabricated by them, conceded that they were probably "marginal tool-makers" (whatever that may mean!).

enging activities of hyaenas. This may have been so in some instances, but Professor Dart has advanced quite formidable arguments against such a thesis. For example, broken up remains of turtles and tortoises and crabs (as well as the egg shells and skulls of a variety of birds) were found in the australopithecine deposits at Taung and Makapansgat and these can hardly be all the dietary products of the predatory or scavenging animals that lived at that time. Reference may be made again to Leakey's first report of the practically complete australopithecine skull found on a living floor at Olduvai. He emphatically states in this report:

> It is of special importance to note that whereas the bones of the larger animals have all been broken and scattered, the hominid skull was found as a single unit within the space of approximately one square foot by about six inches deep. . . . This very great difference between the condition of the hominid skull and that of the animal bones on the same living floor (all of which had been deliberately broken up) seems to indicate clearly that this skull represents one of the hominids who occupied the living site who made and used the tools and who ate the animals. There is no reason, whatever, in this case, to believe that the skull represents the victim of a cannabilistic feast by some hypothetical more advanced type of man. Had we found only fragments of skull, or fragments of jaw, we should not have taken such a positive view of this.

As I have mentioned, Leakey's subsequent discovery of the australopithecine remains to which the name *Homo habilis* was given led him to suppose (mistakenly as I believe) that a more advanced type of hominid did live at Olduvai contemporaneously with the owner of the large and complete australopithecine skull to which he referred in the quotation just given.

The broken animal bones found in such profusion in australopithecine deposits are mostly those of various kinds of antelope, large and small, but they also include at some of the sites a few remains of fossil horses, giraffes, rhinoceroses, warthogs, and baboons, as well as small reptiles. On the Olduvai living floor Leakey has recorded the presence of skeletal fragments of birds, amphibians, snakes, lizards, rodents, antelopes, and specimens of immature fossil pigs. Dart concludes from his examination of the animal bones in the South African deposits that they are generally those of young or old creatures, that is, "those most easily overpowered." The baboon remains are of particular interest, for out of forty-two skulls found at Taung, Sterkfontein and Makapansgat, twenty-seven (64 percent) showed evidence of fractures on the cranial roof [23]. I had the opportunity of examining the baboon skulls from Taung when I visited

Johannesburg in 1946 and it seemed to me that the evidence they presented was highly significant. Many of them showed localized depressed fractures in the parietal region, rather consistently in the same general area of
the skull roof. And in some specimens the fragmented parts of the skull
were but slightly depressed below their original position on the suface. In
other words, as Dart had suggested, they give the appearance of having
resulted from a blow on the head with a clublike weapon of some sort. At
any rate, it is difficult to suppose that depressed fractures of this kind
could have been produced by the teeth of carnivores, or that they could
have been produced accidentally by falls of rock from the roof of caves or
rock shelters. In the latter case the whole skull would surely have been
smashed into bits, and in the former case the sharp fangs of a carnivore
would have produced punctured fractures or the whole skull would have
been scrunched up by the bite of the postcanine teeth. On the other hand,
it is difficult to imagine that the australopithecines would have been
able to attack such formidable creatures as baboons by knocking them on
the head with clubs, particularly as Dart had inferred from the position
of the depressed fractures that many of the blows had been delivered from
the front. But it may be that they were able to attack and kill baboons by
a rather elaborate organization of group hunting. To my mind, however,
the evidence of these fractured skulls, although highly suggestive of attacks by the australopithecines on baboons that they hunted, is not finally
conclusive evidence that the fractures were actually produced in this way.
Unfortunately, fractures of australopithecine skulls and jaws have also
been interpreted as the result of deliberate heavy blows incurred during
life as the result of fights between the australopithecines themselves, and
have even been publicized as the earliest evidence of "murderous" attacks
among those ancient hominids. But it is very difficult to distinguish between fatal injuries that may have been incurred during life, and fractures that may have been produced during the course of fossilization.
After all, many of the australopithecine specimens have lain for many
thousands of years embedded under a heavy weight of many hundreds of
tons of stalagmitic accumulations, and it would be surprising if the pressure of such a heavy weight over such a long time had not led to some degree of breaking of the bones, perhaps long after their initial fossilization.

Professor Dart has made a painstaking analysis of the several thousands of animal bone fragments excavated from bone-breccias in australopithecine deposits, and found that, in the case of the larger animals,
skulls and neck vertebrae were present in far greater numbers than tho-

racic, lumbar, or sacral vertebrae. He therefore drew the conclusion from the selective nature of these bone collections that the heads of these creatures had been severed from the trunk at the site of the kills and carried into the caves or rock shelters. It may also be that, in addition to their hunting activities, the australopithecines were scavengers, in the sense that, as has been reported in some of the modern native African communities, they had the habit of driving away predators in the act of consuming their kills so that they could make use of the remains of the latter for their own consumption. Dart has gone further than this in his interpretation of the animal bone fragments and has suggested that many of these fragments were used by the australopithecines as weapons: the long bones of the limbs as clubs; splintered and sharply pointed bones, antelope horns, and canine tusks as daggers; bones such as the shoulder blade and the jaws of large animals as "slashing" implements; the serrated teeth in the jaws of antelopes as scrapers and sawlike tools. He also maintains [21] that some of the long bone fragments show definite evidence of deliberate shaping for tools by flaking, and of having been used in scraping and rubbing, "dressing" the hides of animals, etc., because of the appearance of smoothed and polished edges or points. He has therefore developed his thesis that the australopithecines had contrived an elaborate "osteodontokeratic" culture, that is, a culture which included a whole armamentarium of tools and weapons constructed of bones, teeth, and horns. He would argue, also, that early hominids are likely to have used such material in this way long before they started to fabricate stone tools and weapons. This contention is not without reason, of course, but it is exceedingly difficult to prove to the satisfaction of archaeologists in general. Dart has certainly demonstrated that bones, teeth, tools, and horns *could* be used as formidable weapons or effective tools, but the evidence that the australopithecines *did* so use them has been seriously questioned. Sharp-pointed bone fragments, it is suggested, may have been the natural result of fractures produced by the pressure of overlying deposits after fossilization or may have been produced by the australopithecines in the course of breaking the long bones to get at the nourishing marrow; the occasional smooth and polished surfaces on some of the fragments may have other causes such as the exposure of these surfaces to lubrication by percolating water or the abrasion of wind-blown sand grains. If a small capuchin monkey can make use of a marrow bone to crush walnuts, it is likely enough, as I have already mentioned, that the australopithecines with their larger brains made use of similar ready-to-hand tools for similar purposes (even though it is not easy to be absolutely certain that some of the broken and fragmented bones found with their remains were actually used in this way). Whether they deliberately shaped bone tools is, of course, another matter. Bone "tools" similar to those found in australopithecine deposits had previously been reported from pithecanthropine sites in China, but even here the interpretation of the bone fragments as artifacts providing evidence of a bone-tool culture has not gained wide agreement among archaeologists. It seems to me that what is needed now is a detailed description and analysis of animal bone fragments in a bone breccia that definitely antedates the evolutionary emergence of the Hominidae, for comparison with those found in australopithecine deposits and analyzed by Dart.

Accepting that the australopithecines were hunters as well as scavengers, the interesting question arises as to their methods of hunting. Presumably they must have used weapons of some sort, whether these were natural objects of bone or stone or deliberately prepared artifacts. But the hunting of animals in open country also implies a social organization of groups of hunters, and the coordination and cohesion of their group activities in the pursuit of game must have depended on a subtle system of communication (whether by vocalization or gesture) between members of each group. This is not to say, of course, that the australopithecines were capable of articulate language; there is no evidence at all that their mental faculties were advanced enough for speech. However, their use of gestures and different vocal sounds to guide each other in their hunting activities may have been the necessary prelude to the development of articulate language in the more highly developed hominids of a much later date. This is about as far as we may safely go in trying to build up a picture of the social organization of the australopithecines without invoking ill-founded speculations.

It is not to be supposed, from what I have said, that the australopithecines were entirely carnivorous in their dietary habits; it may be presumed that they also subsisted on vegetable matter. From a study of the dentition it has been inferred that the robust australopithecines were more vegetarian than the gracile australopithecines, and that the latter were more omnivorous in their diet. This inference is based on the fact that in *Australopithecus robustus* the canine teeth are relatively smaller, and that the molar teeth, being larger, were more adapted for crushing and masticating tough vegetable food; moreover, as I have already mentioned, the molar teeth frequently show small chippings of the enamel at the margins and sides of their occlusal surfaces, similar to the chippings that characterize the molar teeth of creatures like baboons that normally feed on coarse vegetation such as bulbs and roots containing gritty particles. The inference based on this evidence seems fair enough, though in fact it is not always safe to make confident deductions about the food habits of mammals in general from their dental morphology. The modern anthropoid apes, and baboons among the lower primates, normally live on vegetable food. Nevertheless, on occasion they have been observed in the wild to attack and eat small animals, in spite of controversial arguments to the contrary. It may readily be conceded, therefore, that under changing ecological conditions—for example, if communities of higher primates were forced to occupy an open savannah type of country where vegetable food was scarce—normally vegetarian populations would develop their flesh-eating propensities to the extent that they would become predominantly carnivorous. As we have seen, geological evidence makes it clear that the australopithecines did occupy a savannah terrain. The robust australopithecines' more vegetarian diet (compared to the gracile) would accord with the evidence that they lived during a period of increasing rainfall when vegetation might be expected to have become more abundant. This does not exclude the probability, however, that they were also to some extent carnivorous.

It is useful to consider some of the implications arising from the comparative study of the behavior of the modern pongids, for there is perhaps a tendency to assume that because the australopithecine brain was comparable in volume with that of the large apes (though probably larger in proportion to the body weight in the gracile type, *Australopithecus africanus*), the behavioral patterns of the australopithecines were at much the same simian level. But the repertoire of behavioral pattern observed in apes in their natural habitat in the wild appears on the whole to be rather limited. However, it has been pointed out by several observers that the large apes of today inhabit tracts of tropical forest and undergrowth where they are surrounded by an abundance of edible and nutritious vegetable foods, so that they may be said to live a life of almost indolent luxury without the need to elaborate patterns of behavior that would be required to maintain themselves in an environment where a scarcity of vegetable food and perhaps less favorable climatic conditions would call for much greater survival efforts. Now this raises a very interesting point in considering the importance of behavior, and particularly potential behavior, in relation to evolutionary advancement. For it is not what animals *do* do in their natural environment, but what they *can* do under changing and adverse conditions, that will ultimately determine how they will become adapted behaviorally, and finally physically, to new environments. It seems that the idea of "struggle" in the phrase "struggle for existence" is sometimes overlooked, or at least underestimated, animals being regarded as more or less passive material exposed to natural selection by the conditions of the environment in which they find themselves. But progressive evolutionary development surely depends on deliberate efforts made by individual animals, and by the communities of which they are members, to overcome adverse and difficult problems with which they are faced in order to ensure the survival of the species. And, at least in the case of the higher mammals, we may not be far wrong in speaking of their efforts as conscious efforts. If a species by such strivings can manage to survive for a sufficient length of time in surroundings for which it is not as yet fully adapted in its physical makeup, then the opportunity is provided for the gradual development by advantageous mutational variations of morphological changes that adapt the species more perfectly to its ecological environment. Thus the species-environment relationship becomes progressively, and eventually completely, stabilized not only in patterns of behavioral reaction but also in physical adaptations resulting from structural modifications of various kinds. For the elucidation of behavioral potentialities in such animals as the large apes, it is not sufficient to observe them in their wild state (though, of course, this is a most important subject for field studies); a great deal can be learned by observing in captivity their capacity for dealing successfully with new or unusual situations in which they are placed. Everyone knows, for example, that chimpanzees in captivity can be trained to undertake complicated feats such as unlocking a series of locks in their proper sequence in order to get at food inside a box. They can fit bamboo sticks one into another so as to gain a sufficient length for knocking down a banana otherwise out of their reach. They can throw missiles with considerable accuracy of aim. They have been observed to wield clubs apparently with threatening intent. And, of course, they can learn to ride a bicycle, dress and undress, smoke cigarettes, and so forth.

If chimpanzees are capable of complex activities of this kind, some of which evidently involve the preliminary apprehension of a problem to be solved, that is, insight, it can hardly be doubted that the australopithecines were equally capable of dealing with new or unexpected situations arising in their environment, and perhaps more so. The australopithecines occupied an environment that can properly be described as hostile: they lived in a terrain where vegetable food was by no means always abundant; they were exposed in more or less open country of the savannah type where they lacked the natural protection provided by dense forest; they had to defend themselves against large and dangerous animals with weapons of only the crudest kind; they had to face day to day all the risks and uncertainties of primitive foraging or scavenging; they had to hunt game for their food; and (as the geological evidence indicates) they had to contend with fluctuating changes of climate. Clearly, in the face of all these hazards, they must have needed every possible artifice that their wits could devise in their struggle for survival. It may well be, therefore, as Dart has persuasively argued, that they did use tools and weapons of animal bones, horns, and teeth in the manner he suggests, even if no objective proof of such activities can be adduced to satisfy archaeologists. But the australopithecines had one outstanding advantage; they had achieved an erect bipedalism that enabled them to run with speed and to use their hands freely for manipulative purposes. The erect bipedalism of early hominids was no doubt of primary importance for the subsequent evolutionary development of more advanced hominids. The view was at one time widely held that it was the preliminary expansion of the brain that conferred on man the ability to fabricate tools and weapons of stone or other material. It now seems probable that it was a reverse process, that it was the ability to construct implements, however crude, which provided the stimulus for the evolutionary development of larger, and still larger, brains. It is obvious enough that individuals and communities endowed with larger brains and more acute mental faculties enabling them to manufacture more effective implements for attack and defense, or for necessary requirements like the collection and preparation of food, and so forth, would have an advantage over other individuals and communities less well endowed. In other words, it is here a matter of natural selection permitting the spread of favorable mutations among the australopithecine populations.

The same line of reasoning leads to the supposition that in the course of evolutionary development from the australopithecine phase of the Early Pleistocene, the structure of the hand became gradually more refined to permit greater dexterity and more delicate manipulative activities, and in the course of thousands of years the necessary anatomical basis for erect bipedalism gradually became modified to provide a much more efficient mechanism for walking and running. We may suppose, indeed, that the australopithecines were engaged in a struggle for existence with a physical equipment of anatomical structure not as yet fully adapted to the needs of their environment. But by conscious efforts to overcome the difficulties of their environment by sheer ingenuity and improvisation, they did manage to survive and thus to allow time for the exploitation by natural selection of such adaptive bodily mutations that might occasionally appear in individuals. Adaptive modifications of this sort, as the result of their survival value under strong selective pressures, would be expected to spread by slow degrees through the population as a whole. It is in such a way that the transition from the australopithecine to the pithecanthropine phase of hominid evolution can be envisaged. The pithecanthropines owed their origin to the unremitting struggles of the australopithecines for survival. And in the same way modern *Homo sapiens* owes his origin to the vigorous efforts of the pithecanthropines to maintain their existence in the face of the incredibly difficult situations with which they must frequently have had to cope.

10

EVOLUTIONARY ORIGINS
OF THE AUSTRALOPITHECINES

Taxonomically speaking, the affinities of the genus *Australopithecus* with the genus *Homo* are very close. Indeed, on the basis of its morphological characters, a taxonomist of repute suggested at one time that it should actually be included in *Homo* and be given no more than specific rank to distinguish it from the pithecanthropines, *Homo erectus*. As we have seen, on dental morphology alone it may be difficult to differentiate between the smaller australopithecines and some of the pithecanthropines. However, there are quite marked contrasts in brain size and in limb structure between *Australopithecus* and *Homo* that are sufficient to demand the generic distinction which this nomenclature implies, and this distinction is now widely agreed upon by all students of fossil primates. It remains true that the two genera are taxonomically far more closely related to each other than either is to any of the known representatives of the Pongidae. It is also the case that the main gap in the fossil record of hominid evolution is to be found not between the australopithecines and the pithecanthropines, but between the australopithecines of the Early Pleistocene and the extinct apelike creatures whose remains have been

125

discovered in deposits laid down at a much earlier date in the geological periods of the Pliocene and Miocene. Many of these remains have been found, consisting unfortunately in most cases of no more than teeth and fragments of jaws, but a notable exception is the discovery by Dr. Leakey of a skull and a number of limb bones of a Miocene ape that once inhabited East Africa in the region of Lake Victoria.

More than twenty genera of extinct apes have been given different names by the palaeontologists who have from time to time described these fossil remains, but the distinctions that have provided the excuse for this exuberant nomenclature have usually been based on rather trivial and unimportant characters of the teeth. The fact is that in the past we have tended to become too "hypnotized" by small variations in tooth structure, in spite of the fact that, as long ago as 1922, it was demonstrated by Remane [63] that individual variations in the dentition of a single species of the modern anthropoid apes are very considerable and they may far exceed the small differences on which generic distinctions among fossil apes have been claimed. Quite recently a drastic, but highly important and useful, revision of the complicated nomenclature of the large Miocene and Pliocene apes has been completed by Professor Simons and Dr. Pilbeam [82]. In place of the twenty odd generic names scattered through the literature of the subject, they recognize no more than three genera, *Dryopithecus, Gigantopithecus,* and *Ramapithecus,* of which the first contains a number of separate species. This is a most welcome simplification of the nomenclature of the fossil apes, even though it is likely that some palaeontologists may regard it as an oversimplification.

The large fossil apes of Miocene age from East Africa are of particular importance, for they clearly demonstrate that these creatures were much more generalized in their anatomical structure than the large modern anthropoid apes.

They were initially given the generic name *Proconsul,* a name which has a certain air of flippancy about it because there was once a chimpanzee called Consul that used to appear on the music-hall stage and was famed for the elaborate tricks it had been trained to perform; *Proconsul* was the implied ancestor of the modern chimpanzee (which may well have been the case). The simplified nomenclature of Simons and Pilbeam has now placed these East African fossils in the genus *Dryopithecus,* recognizing three species, *D. africanus, D. nyanzae,* and *D. major,* these species varying in size from that of the pigmy chimpanzee of today to a large gorilla. We know most about the first, and smallest, of the three species;

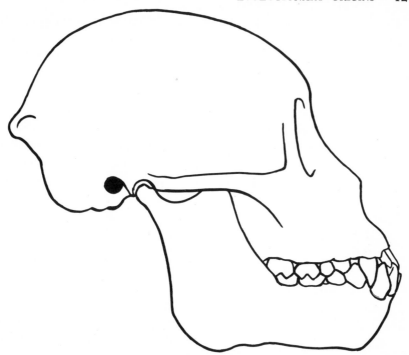

Figure 33. The skull of *Dryopithecus (Proconsul) africanus* as reconstructed by Dr. P. R. Davis and Dr. J. R. Napier. (Reprinted from *The Fossil Evidence for Human Evolution*, 2d ed., by Sir Wilfrid E. Le Gros Clark, © 1955 by the University of Chicago, by permission of the University of Chicago Press.)

Dr. Leakey not only found its skull (incidentally the only skull of a Miocene ape so far discovered), but also some of the arm and leg bones. The skull is lightly built with no protruding brow ridges and the facial skeleton is relatively short (see Fig. 33). The canine teeth overlap in occlusion and are strongly projecting, and the front lower premolar tooth has a sectorial character. On the other hand, the incisor teeth are not enlarged as they are in the modern apes. In the upper limb skeleton the forearm is relatively short and the thumb relatively long; in other words, the upper limb shows none of the extreme specializations developed in the modern apes for brachiating habits. The thighbone is slender in build, and two of the anklebones, the heel bone, or calcaneus, and the talus show certain resemblances to those of the Old World monkeys. It has been inferred from such characters of the limb bones that *Dryopithecus africanus* was not specialized for a completely arboreal habitat, but, while it may have been an active climber, it was also able to scamper along the ground like many of the quadrupedal monkeys of today. So generalized in structure were the limb bones that this Miocene ape might well be a representative of the ancestral stock that gave rise by divergent modifications during later geological times to the modern pongids in one direction, and to the hominids in another. That the adoption of bipedal terrestrial habits by the earliest representatives of the Hominidae must have occurred in regions of deforestation (the result of changing climatic conditions) seems very probable; further, the environment of the East African Miocene apes evidently provided possibilities of this sort, as indicated by a consideration of the fossilized fruits, seeds, and insects discovered by Dr. Leakey in the contemporary deposits. I have elsewhere [50] made the following suggestion:

> . . . the evolution of ground-living forms in the ancestry of the Hominidae was the result of adaptations primarily concerned, not with the abandonment of arboreal life, but (paradoxically) with an attempt to retain it. For in regions undergoing gradual deforestation they would make it possible to cross intervening grasslands in order to pass from one restricted and shrinking wooded area to another. This proposition is parallel to the interesting conjecture that water-living vertebrates initially acquired terrestrial and air-breathing adaptations in order to preserve their aquatic mode of life; for in times of drought these adaptations would make it possible to escape from dried-up rivers or pools and go overland in search of water elsewhere.

It is unfortunate that we know nothing of the limb skeleton of the dryopithecines that lived elsewhere than in East Africa, except only for a humerus and a femur found in Europe, the former very incomplete. But they are both slender and lightly built bones similar to those of *Dryopithecus africanus*. The European dryopithecine jaws and teeth are also somewhat similar, particularly in the generalized structure of the molar teeth, but they show also certain differences in minor characters. No doubt the most important fossil "ape" that has a bearing on hominid evolution is represented by the genus *Ramapithecus* of Upper Miocene or Lower Pliocene date, say ten to fifteen million years ago. I have enclosed "ape" in quotation marks because it is a matter of discussion whether it really is an ape in the sense of belonging to the evolutionary lineage of the Pongidae. On the contrary, it has been argued on the basis of the several fragmentary specimens so far discovered that it is a very primitive member of the Hominidae. If this is really so, it is the only hominid known whose remains date from pre-Pleistocene times.

The most detailed studies of *Ramapithecus* are those of Professor Simons [78]. The fossil remains indicate that it had a wide geographical range, from India to East Africa, and perhaps as far east as China. The East African fossil consists of an upper jaw with much of the upper dentition found by Leakey in Kenya in deposits of approximately the same age as those which yielded *Ramapithecus* specimens in India. Leakey named his specimen *Kenyapithecus* on the basis of trivial differences from the Indian fossils, but, as Simons has observed [80] "greater differences than have been noted here typically occur among members of a single family social group within nearly all species of present day hominoids." The evidence so far available shows that *Ramapithecus* was characterized by a parabolic dental arcade; small incisors, canines, and premolars; no pronounced gap (diastema) between the upper canine tooth and the adjacent incisor tooth; and a relatively short or orthognathous facial skeleton. Simons and Pilbeam further remark that "the shallow robust mandible differs from that of *Dryopithecus* and recalls later hominids such as *Australopithecus*," and also that, since the East African dryopithecines probably range back to the Middle Miocene or even earlier, *Ramapithecus* may have had its evolutionary origin from a primitive species of *Dryopithecus*. Whether the shortened face and the small size of the front teeth of *Ramapithecus* are to be correlated with an incipiest bipedalism and the development of tool-using propensities is a matter for conjecture; it is an interesting suggestion but one that is not as yet supported by objective evidence. The geographical distribution of *Ramapithecus* clearly indicates that it was a very wide-ranging species, and it is, of course, possible that its mobility over such far distances in the Old World may have been facilitated by bipedalism.

We may complete our brief reference to fossil apes by noting that *Gigantopithecus,* a genus that inhabited China during the Pliocene, is regarded as a side branch of the main dryopithecine group, differing little from *Dryopithecus* except in the size of its jaws and teeth.

So far, then, there is a reasonably good fossil record of hominid phylogeny leading in a graded morphological and temporal sequence from the primitive and generalized hominoid genus *Dryopithecus* (that presumably provided a common ancestry for the subsequent evolutionary divergence of the two families Pongidae and Hominidae) to the earliest hominids represented perhaps by *Ramapithecus,* to *Australopithecus,* to *Homo erectus,* and finally to *Homo sapiens.* The question arises whether, as some have argued to be the case, there is any theoretical objection to postulating the derivation of the hominid type of dentition from the more pongidlike type of dentition of *Dryopithecus*. It has been contended, for example, that the projecting conical canine teeth and the sectorial lower front premolars found in this genus (wherein it approximates closely to the modern pongids) are to be regarded as "specializations" that could not have found a place in hominid ancestry. But this line of argument has been based on the assumption that these are "specializations" that were not capable of undergoing an evolutionary reversal, and such a premise is certainly without foundation. In the first place it may be questioned whether they are properly to be regarded as "specializations," and even if this view is correct the subsequent evolutionary reduction of the canine tooth and the conversion of a sectorial premolar into a bicuspid tooth can readily be postulated without doing violence to any of the known genetical principles involved in evolutionary change. So far as the canine tooth is concerned, even in the genus *Dryopithecus* it actually showed a good deal of variation in its relative size. But there is also indirect evidence that the small spatulate canine of modern man is the result of a secondary reduction in size during the course of evolution. For example, as I have previously mentioned, the newly erupted tooth may be quite sharply pointed and project well beyond the level of the adjacent teeth, and these characters may be obvious enough in some individuals until, in the early stages of attrition, the tip of the canine becomes worn down to a flattened surface on a level with the adjacent teeth. Again, the canine tooth of *Homo sapiens* is provided with an unusually robust root which is also longer than the roots of the incisor and premolar teeth; such features are difficult to explain on a purely functional basis, for in modern man the canine teeth have no special function to perform. But they do become intelligible if we suppose that they had special functions in the past. There is also the evidence of fossil hominids that the modern canine tooth has undergone retrogressive modifications, for in some of the pithecanthropines of half a million years ago this tooth was much larger and more powerfully constructed. The sectorial character of the front lower premolar tooth in the extinct hominoid genus *Dryopithecus* and in pongids generally is little more than a functional correlate of long projecting and interlocking canines, for it is functionally related to the shearing action of the upper canine in biting movements. It may be confidently assumed, therefore, that a reduction of the canine to a spatulate form would be accompanied by a modification of the opposing lower premolar tooth with a loss of its sectorial character, and there is therefore no difficulty in explaining its conversion into the bicuspid type of tooth that is characteristic of the Hominidae.

To complete the story of human evolution, with reference to the origin of the early hominids such as *Australopithecus* in particular, and to the ultimate origin of man in general, we may refer briefly to the antecedents of the dryopithecines of Miocene and Pliocene times. These antecedents are known from Oligocene deposits in Egypt, deposits that date back perhaps thirty or forty million years. Over half a century ago the lower jaw of a small monkeylike Oligocene primate, with its full complement of teeth, was found at this site. It was given the name *Parapithecus*. This tiny creature—about the size of the little squirrel monkey of today—is certainly one of the higher primates, but whether it should be classified as an Old World monkey or as an exceedingly primitive kind of anthropoid ape is a matter for argument. Perhaps it was neither in the strict taxonomic sense, but a generalized ancestral type from which, following divergent evolutionary trends, both the monkeys and the apes were derived. In certain features of its dentition and also in the conformation of its mandible, *Parapithecus* shows some most interesting resemblances to a group of lower primates called tarsioids of which only one representative remains extant at the present time—the little tarsier that inhabits the jungles of Borneo, the Philippines and the Celebes. But during the geological period called the Eocene, which preceded the Oligocene period, many different genera of tarsioids were in existence. It has been inferred from their fossil remains that one or other group of these extinct tarsioids probably gave rise to the higher primates, that is, to the monkeys and apes. Hence *Parapithecus* is of particular importance, for it seems to represent a transitional stage between the lower and the higher primates. Also found more recently in the Oligocene deposits of Egypt was a small primate, discovered and described by Professor Simons and named by him *Oligopithecus* [79]. Again it is uncertain whether this was a primitive anthropoid ape or an early representative of the evolutionary line leading to the Old World monkeys; the latter interpretation is, perhaps, the more likely. Like *Parapithecus,* but to a lesser degree, it shows certain details of dental morphology that suggest tarsioid affinities. Simons found in the same locality the frontal part of a skull, up to now the oldest skull fragment of a higher primate yet known [76]. It quite closely resembles in size and shape the frontal part of the skull of the little marmosets of the New World and possibly was part of the skull of *Oligopithecus*. Besides these extremely primitive genera, another that is more definitely to be classified as an early anthropoid ape (rather than a monkey) is also known from the same Egyptian deposits. This is called *Propliopithecus* because it was thought to be ancestral to a fossil gibbon-like ape, *Pliopithecus,* which was common in Europe, Africa, and elsewhere during Miocene and Pliocene times, and, in its turn, *Pliopithecus* has been taken to represent the ancestor of the modern gibbons which today occupy tropical forests in the Far East. But the evidence so far available does not allow one to be dogmatic about such assumptions. For example, *Propliopithecus* may, with equal probability, represent the ancestral stock from which the dryopithecines were derived. Here is another gap in the fossil record of the higher primates that is waiting to be filled, and to fill it with some degree of assurance we need more than teeth and fragments of jaws—we need reasonably good specimens of skulls and limb bones of the Oligocene primates. Doubtless these will come to hand in time, particularly if the fossil-collecting expeditions to Egypt by Professor Simons are continued and extended.

The fossil record of the higher primates preceding the australopithecines is still far from complete, and for this reason, of course, interpretations based on it can be no more than provisional interpretations. But the record is becoming more and more complete, year by year, and the time may not be far distant when the fossil evidence will become sufficiently adequate to permit conclusions regarding the early stages of hominoid evolution capable of being drawn with much greater assurance than is at present possible.

POSTLUDE

This book has been concerned with the story of the discoveries in Africa of the fossil remains of the australopithecines, the small-brained hominids that lived there a million years ago or more. In spite of their small brains, they had already acquired, in a form as yet imperfect, the anatomical requirements for an erect bipedal posture and gait, and for a degree of manual dexterity that made it possible for them to use or fabricate tools and weapons. Thus they foreshadowed the more advanced development of later hominids such as the pithecanthropines, and eventually of the species *Homo sapiens* of which we ourselves are members. There can now be little doubt that, as a group, they were ancestral to these later types. These have been very remarkable discoveries and it is perhaps likely that we might not know anything about them today but for the indefatigable explorations of three outstanding field workers, Professor Raymond Dart, Dr. Louis Leakey, and the late Dr. Robert Broom. It is primarily because of them and their associates that so many remains of the australopithecines have been brought to light. Their reports of these discoveries at first aroused a good deal of adverse criticism and in some cases led to polemical controversies of a nature not altogether appropriate to scientific discussions, nor conducive to a dispassionate appraisal of the evidence. These controversies have now faded into the background, as the progressive accumulation of more and more fossil australopithecine teeth and bones of the postcranial skeleton has served to substantiate, in general, the conclusions about their fundamental significance made by their original discoverers. This is not to say, of course, that without exception all their interpretations of the australopithecine material have been accepted by other palaeontologists or have been proved correct; basically, however, their main conclusions have been validated by careful and systematic studies of the anatomical evidence. It may be that some of the claims they have made are open to question or have been invalidated, but what matter if this be so? Probably no scientist, however distinguished, has been invariably correct in the interpretations and theoretical hypotheses that he has advanced to explain his observational or experimental data. Hypotheses are of vital importance for progress in all branches of scientific knowledge, for only by formulating hypotheses is it possible to put them to the test by further observation and experiment, or by the application of new and more refined techniques. It is perhaps the fate of many hypotheses to be superseded sooner or later. If they are genuinely scientific hypotheses, that is to say, if they are hypotheses that accord reasonably well with the facts available to their authors *at the time* and are susceptible to the test of observation and experiment, they will have played their part in promoting and accelerating the advance of scientific thought and research even if eventually they are discarded. A well-known Scottish author, Samuel Smiles, once remarked, over a hundred years ago: "Probably he who never made a mistake never made a discovery." There is a great deal of truth in this aphorism, and certainly those who have made so many and such important discoveries by their energetic field work readily may be excused if one or other of their hypotheses should eventually prove to be unsubstantiated or untenable. Their positive contributions to the study of human evolution immeasurably outweigh such of their provisional interpretations as may be negated by more detailed inquiries or by future discoveries.

SELECTED REFERENCES

(1) Abel, W. "Kritische Untersuchungen über *Australopithecus africanus,*" *Morphol. Jahrb.,* **65** (1931), 541.

(2) Bartholomew, G., A., and J. B. Birdsell. "Ecology and the Proto-hominids," *Am. Anthropol.,* **55** (1953), 481.

(3) Brain, C. K. "The Transvaal Ape-man-Bearing Deposits," *Transvaal Museum Mem.,* No. 11 (1958).

(4) Bronowski, J., and N. M. Long. "Statistical Methods in Anthropology," *Nature,* **168** (1951), 794.

(5) Broom, R. "Discovery of a New Skull of the South African Ape-man, *Plesianthropus,*" *Nature,* **159** (1947), 672.

(6) ———. "The Mandible of the Sterkfontein Ape-man, *Plesianthropus,*" *South African Sci.,* **1** (1947), 14.

(7) ———, and J. T. Robinson. "Further Remains of the Sterkfontein Ape-man, *Plesianthropus,*" *Nature,* **160** (1947), 430.

(8) ———. "The Lower End of the Femur of *Plesianthropus,*" *Ann. Transvaal Museum,* **21** (1949), 181.

(9) ———. "Thumb of the Swartkrans Ape-man," *Nature,* **164** (1949), 841.

(10) ———. "A New Mandible of the Ape-Man, *Plesianthropus transvaalensis,*" *Am. J. Phys. Anthropol.,* **7** (1949), 123.

(11) ———. "A New Type of Fossil Man," *Nature,* **164** (1949), 322.

(12) ———. "Swartkrans Ape-man," *Transvaal Museum Mem.,* No. 6 (1952).

(13) ———, J. T. Robinson, and G. W. H. Schepers. "Sterkfontein Ape-man, *Plesianthropus,*" *Transvaal Museum Mem.,* No. 4 (1950).

137

(14) ——, and G. W. H. Schepers. "The South African Fossil Ape-man: The Australopithecinae," *Transvaal Museum Mem.,* No. 2 (1946).

(15) Coppens, Y. "Découverte d'un Australopitheciné dans le Villanfranchien du Tchad," *Problems Actuels de Paléontologie* (Paris), p. 455.

(16) Dart, R. A. "*Australopithecus africanus:* The Man-ape of South Africa," *Nature* (Feb. 7, 1925), p. 1.

(17) ——. "An Adolescent Promethean Australopithecine Mandible from Makapansgat," *South African Sci.,* 2 (1948), 73.

(18) ——. "The First Pelvic Bones of *Australopithecus prometheus,*" *Am. J. Phys. Anthropol.,* 7 (1949), 301.

(19) ——. "Innominate Fragments of *Australopithecus prometheus,*"* *Am. J. Phys. Anthropol.,* 7 (1949), 301.

(20) ——. "*Australopithecus prometheus* and *Telanthropus capensis,*" *Am. J. Phys. Anthropol.,* 13 (1955), 67.

(21) ——. "The Osteodontokeratic Culture of *Australopithecus prometheus,*" *Transvaal Museum Mem.,* No. 10 (1957).

(22) ——. "A Further Adolescent Ilium from Makapansgat," *Am. J. Phys. Anthropol.,* 16 (1958), 473.

(23) ——. *Adventures with the Missing Link.* London: Hamish Hami'ton, 1959.

(24) ——. "The First *Australopithecus* Cranium from the Pink Breccia at Makapansgat," *Am. J. Phys. Anthropol.,* 17 (1959), 77.

(25) ——. "A Cleft Adult Mandible and Nine Other Lower Jaw Fragments from Makapansgat," *Am. J. Phys. Anthropol.,* 20 (1962), 267.

(26) Davis, P. R. "Hominid Fossils from Bed 1, Olduvai Gorge, Tanganyika: A Tibia and Fibula," *Nature,* 201 (1964), 967.

(27) Day, M. H., and J. R. Napier. "Hominid Fossils from Bed 1, Olduvai Gorge, Tanganyika: Fossil Foot Bones," *Nature,* 201 (1964), 969.

(28) Dobzhansky, T. *Mankind Evolving.* New Haven, Conn.: Yale University Press, 1962.

(29) Ewer, R. F. "Faunal Evidence on the Dating of the Australopithecinae," *Proc. Third Pan-African Congr. on Prehistory* (1955), p. 135.

(30) Goodall, J. "Feeding Behavior of Wild Chimpanzees," *Symp. Zool. Soc. London,* No. 10 (1963), 39.

(31) Goodman, M. "Man's Place in the Phylogeny of the Primates as Reflected in Serum Proteins," in *Classification and Human Evolution,* ed. S. L. Washburn. London: Methuen & Co., 1964, p. 204.

(32) Gregory, W. K. "The Origin of Man from a Brachiating Anthropoid Stock," *Science,* 71 (1930), 645.

(33) ——. "The South African Fossil Man-apes and the Origin of the Human Dentition," *J. Am. Dent. Assoc.,* 26 (1939), 288.

(34) ——, and M. Hellman. "The Dentition of the Extinct South African Man-ape *Australopithecus transvaalensis:* A Comparative and Phylogenetic Study," *Ann. Transvaal Museum,* 19 (1939), 339.

* Note that the fossil remains initially designated as *Australopithecus prometheus* are now generally recognized to be those of *Australopithecus africanus.*

(35) Haldane, J. B. S. "Can a Species Concept be Justified?" in *The Species Concept in Palaeontology,* Systematics Assoc. Publ., No. 2, ed. P. C. Sylvester. London: Bradley, 1956.

(36) Harrison, G. A., and J. S. Weiner. "Some Considerations in the Formulation of Theories of Human Phylogeny," in *Classification and Human Evolution,* ed. S. L. Washburn. London: Methuen & Co., 1963, p. 75.

(37) Keith, A. "Australopithecinae or Dartians?" *Nature,* 159 (1947), 377.

(38) Kern, H. M. and W. L. Straus. "The Femur of *Plesianthropus transvaalensis,*" *Am. J. Phys. Anthropol.,* 7 (1949), 53.

(39) Klinger, H. P., J. L. Hamerton, D. Mutton, and E. M. Lang "The Chromosomes of the Hominoidea," in *Classification of Human Evolution,* ed. S. L. Washburn. London: Methuen & Co., 1964, p. 235.

(40) Krogman, W. M. "Studies on Growth Changes in the Skull and Face of Anthropoids," *Am. J. Anat.,* 46 (1930), 315.

(41) Leakey, L. S. B. "A New Fossil Skull from Olduvai," *Nature,* 184 (1959), 491.

(42) ———. "Recent Discoveries at Olduvai Gorge," *Nature,* 188 (1960), 1050.

(43) ———, and M. D. Leakey. "Recent Discoveries of Fossil Hominids in Tanganyika at Olduvai and near Lake Natron," *Nature,* 202 (1964), 5.

(44) ———, P. V. Tobias, and J. R. Napier. "A New Species of the Genus *Homo* from Olduvai Gorge," *Nature,* 202 (1964), 7.

(45) Le Gros Clark, W. E. "The Interpretation of Human Fossils," *Mod. Quart.,* 2 (1939), 115.

(46) ———. "Palaeontological Evidence Bearing on Human Evolution," *Biol. Rev.,* 15 (1940), 202.

(47) ———. "Observations on the Anatomy of the Fossil Australopithecinae," *J. Anat.,* 81 (1947), 300.

(48) ———. "Hominid Characters of the Australopithecine Dentition," *J. Roy. Anthropol. Inst.,* 58 (1952), 37.

(49) ———. *The Antecedents of Man,* 2nd edition. Edinburgh University Press, 1962.

(50) ———. *The Fossil Evidence for Human Evolution,* 2nd edition. Chicago University Press, 1964.

(51) ———. *History of the Primates,* 9th edition. London: British Museum (Natural History), 1965.

(52) Mason, R. J. "The Sterkfontein Stone Artifacts and their Maker," *South African Archaeol. Bull.,* 17 (1962), 109.

(53) Napier, J. R. *Fossil Metacarpals from Swartkrans. Fossil Mammals of Africa,* No. 17. London: British Museum (Natural History), 1959.

(54) ———. "Fossil Hand Bones from Olduvai Gorge," *Nature,* 196 (1962), 429.

(55) ———. "Profile of Early Man at Olduvai," *New Sci.,* No. 386 (1964), p. 80.

(56) ———. "The Evolution of Bipedal Walking in the Hominids," *Archaeol. Biol.,* 75 (1964), 673.

(57) Oakley, K. P. "The Dating of the Australopithecinae of Africa," *Am. J. Phys. Anthropol.,* 12 (1954), 9.

(58) ———. "Study Tour of Early Hominid Sites in Southern Africa," *South African Archaeol. Bull.,* 9 (1954), 75.

(59) ———. "Evidence of Fire in South African Cave Deposits," *Nature,* 174 (1954), 261.

(60) ———. "The Earliest Tool Makers," in *Evolution and Hominization.* Stuttgart: G. Fischer, 1962, p. 157.

(61) ———. *Frameworks for Dating Fossil Man.* London: Weidenfeld and Nicolson, 1964.

(62) Pilbeam, D. R.. and E. L. Simons. "Some Problems of Hominid Classification," *Am. Sci.,* 53 (1965), 337.

(63) Remane, A. "Beiträge zur Morphologie des Anthropoidengebisses," *Archaeol. Naturgesch.,* 87 (1922), 1.

(64) Robinson, J. T. *"Telanthropus* and its Phylogenetic Significance," *Am. J. Phys. Anthropol.,* 11 (1953), 445.

(65) ———. *"Meganthropus,* Australopithecines, and Hominids," *Am. J. Phys. Anthropol.,* 11 (1953), 1.

(66) ———. "The Dentition of the Australopithecinae," *Transvaal Museum Mem.,* No. 9 (1956).

(67) ———. "Occurrence of Stone Artifacts with *Australopithecus* at Sterkfontein," *Nature,* 180 (1957), 521.

(68) ———. "Cranial Cresting Patterns and their Significance in the Hominoidea," *Am. J. Phys. Anthropol.,* 16 (1958), 397.

(69) ———. "The Affinities of the New Olduvai Australopithecine," *Nature,* 186 (1960), 456.

(70) ———. "Sterkfontein Stratigraphy and the Significance of the Extension Site," *South African Archaeol. Bull.,* 17 (1962), 87.

(71) ———. "New Discoveries in Tanganyika: Their Bearing on Hominid Evolution," *Current Anthropol.,* 6 (1965), 403.

(72) ———. *"Homo 'habilis'* and the Australopithecines," *Nature,* 205 (1965), 121.

(73) Schultz, A. H. "Eruption and Decay of the Permanent Teeth in Primates," *Am. J. Phys. Anthropol.,* 19 (1935), 489.

(74) ———. "Age Changes, Sex Differences, and Variability as Factors in the Classification of the Primates," in *Classification and Human Evolution,* ed. S. L. Washburn. London: Methuen & Co., 1964, p. 85.

(75) Schuman, E. L., and C. L. Bruce. "Metric and Morphologic Variations in the Dentition of the Liberian Chimpanzee," *Human Biol.,* 26 (1954), 239.

(76) Simons, E. L. "An Anthropoid Frontal Bone from the Fayum Oligocene of Egypt," *Am. Museum Novit.,* No. 1976 (1959).

(77) ———. "An Anthropoid Mandible from the Oligocene Fayum Beds of Egypt," *Am. Museum Novit.,* No. 2051 (1961).

(78) ———. "The Phyletic Position of *Ramapithecus," Postilla,* Yale Peabody Museum. No. 57 (1961).

(79) ———. "Two New Primate Species from the African Oligocene," *Postilla,* Yale Peabody Museum, No. 64 (1962).

(80) ———. "Some Fallacies in the Study of Human Phylogeny," *Science*, **141** (1963), 879.

(81) ———. "New Fossil Apes from Egypt and the Initial Differentiation of the Hominoidea," *Nature*, **205** (1965), 135.

(82) ———, and D. R. Pilbeam. "Preliminary Revision of the Dryopithecinae," *Folia Primatologia*, **3** (1965), 81.

(83) Simpson, G. G. *Horses*. London: Oxford University Press, 1951.

(84) Stekelis, M., L. Picard, N. Schulman, and G. Haas. "Villafranchian Deposits near Ubeidya in the Central Jordan Valley," *Bull. Res. Council Israel*, **9** (1960), 175.

(85) Straus, W. L. "The Humerus of *Paranthropus robustus*," *Am. J. Phys. Anthropol.*, **6** (1948), 285.

(86) Symington, J. "Endocranial Casts and Brain Form: Criticism of Some Recent Speculations," *J. Anat. Physiol.*, **50** (1916), 111.

(87) Thomas, G. "The Species Conflict: Abstractions and their Applicability," in *The Species Concept in Palaeontology*, Publ. No. 2 of the Systematics Assoc., London (1956).

(88) Tobias, P. V. "Cranial Capacity of *Zinjanthropus* and other Australopithecines," *Nature*, **197** (1963), 743.

(89) ———. "The Olduvai Bed 1 Hominine with Special Reference to its Cranial Capacity," *Nature*, **202** (1964), 3.

(90) ———. "Early Man in East Africa," *Science*, **149** (1965), 22.

(91) Vevers, G. M., and J. S. Weiner. "Use of a Tool by a Captive Capuchin Monkey," *Symp. Zool. Soc. London*, No. 10 (1963), 115.

(92) Washburn, S. L. "Australopithecines: The Hunter or the Hunted?" *Am. Anthropol.*, **59** (1957), 612.

(93) Yates, F., and M. J. R. Healey. "Statistical Methods in Anthropology," *Nature*, **168** (1951), 116.

(94) Zuckerman, S. "South African Anthropoids," *Nature*, **166** (1950), 188.

(95) ———. "Correlation of Change in the Evolution of Higher Primates," in *Evolution as a Process*, ed. J. Huxley and E. B. Ford. London: George Allen and Unwin, 1954, p. 300.

INDEX

A

Abel, Wolfgang, 26
Aboriginal skull, 7*
Absolute dating methods, 11, 12–13, 44
Acetabular socket, 89, 91, *92*
Anatomical characters, comparison of
   patterns, 35
   of independent acquisition, 23, 25,
      26
   of independent requisition, *24*
   of Pongidae and Hominidae, 23–25,
      24
   taxonomic relevance in, 24, 25, 37–
      38, 42
Anthropoid apes, 1, 2
   anklebones of, *100*, 101
   blood groups of, 4
   brain of, 3, 4
   genetic relationship of to hominids,
      5
   Homo sapiens compared with, *88*
   limb bones of, 3
      See also Limb bones
   orthograde posture of, 3
   parasitic infestation of, 4
   pelvis of, 87–91, *88, 90, 95*
   skull of, 3, 7, 66–68, *67,* 70–71, 74,
      76, 77, 79
   teeth of, 3–4, *52,* 53–57, *54*
Anthropomorpha, 1

Ape-men, 2
Atlanthropus mauritanicus, 13
Australopithecine limb bones, 28–29,
      43
   of ankle, 99–101, *100*
   and elbow joint hyperextension, 99
   of foot, 101–102, *102*
   of hand, 43–44, 106–110, *108*
   of lower leg, 102–103
   of thigh, 34, 103–106, *104*
   of upper arm, 98–99
Australopithecine pelvis, 43, 87–97, *88*
   hominid configuration of, 91, 94
Australopithecine skull, 15–16, 18–19,
      20–22, 28, 30, 65–66, *75, 77*
   and basicranial axis, 75, 78, *79*
   controversy over, 32–33
   dolichocephalic nature of, 21
   foramen magnum of, *76, 77, 79*
   fusion of cranial vault sutures in,
      82, *83*
   glenoid fossa of, 81
   gorilla skull compared with, *76, 77,
      79*
   hominid nature of, 74–82, *79*
   immature, *71, 72*
   infra-orbital foramen of, 80
   jaw of (*see* Jaw)
   mastoid process in, *75,* 81, *83*
   orbital aperture of, 80
   "qualitative" features of, 84–86

* Page numbers in italics refer to illustrations.

145